K

The Future of Post-Human
Space-Time

PETER LANG
New York • Washington, D.C./Baltimore • Bern
Frankfurt am Main • Berlin • Brussels • Vienna • Oxford

Peter Baofu

The Future of Post-Human Space-Time

Conceiving a Better Way to Understand Space and Time

PETER LANG
New York • Washington, D.C./Baltimore • Bern
Frankfurt am Main • Berlin • Brussels • Vienna • Oxford

Library of Congress Cataloging-in-Publication Data
Baofu, Peter.
The future of post-human space-time: conceiving a better way
to understand space and time / Peter Baofu.
p. cm.
Includes bibliographical references and index.
1. Space and time. I. Title.
BD632.B36   115—dc22   2006022859
ISBN 0-8204-8871-2

Bibliographic information published by **Die Deutsche Bibliothek**.
Die Deutsche Bibliothek lists this publication in the "Deutsche
Nationalbibliografie"; detailed bibliographic data is available
on the Internet at http://dnb.ddb.de/.

Cover design by Joni Holst

The paper in this book meets the guidelines for permanence and durability
of the Committee on Production Guidelines for Book Longevity
of the Council of Library Resources.

© 2006 Peter Lang Publishing, Inc., New York
29 Broadway, New York, NY 10006
www.peterlang.com

All rights reserved.
Reprint or reproduction, even partially, in all forms such as microfilm,
xerography, microfiche, microcard, and offset strictly prohibited.

Printed in Germany

To Intelligent Life in Post-Human Space-Time,
Long After the End of Human Space-Time

Contents

List of Tables..*ix*
Foreword (Sylvan von Burg)....................................... *xiii*
Acknowledgments ....*xv*
List of Abbreviations ....*xvii*

Part One: Introduction

Chapter One. Introduction: Space-Time and Humans 3
    Space-Time in Classical Mechanics and
        the Childhood of Science 3
    Space-Time in the Theory of Relativity and
        the Adolescence of Science 5
    The Perspectival Theory of Space-Time 7
    The Nature of Existential Dialectics 8
    Methodological Holism..14
    Chapter Outline 15
    Previous Works, Book Title, and Neologisms........................16

Part Two: Culture

Chapter Two. Space-Time and Culture.. 95
    Space-Time and the Influence of Culture 95
    Space and Culture 96
    Time and Culture 101
    Space-Time and the Delimitation of Culture........................107

Part Three: Society

Chapter Three. Space-Time and Society..111
    Space-Time and the Power of Society...................................111
    Space-Time and Social Organizations 112
    Space-Time and Social Institutions...................................... 116
    Space-Time and Social Structure...120
    Space-Time and Social Systems..124
    Space-Time and the Limits of Society 131

Part Four: The Mind

Chapter Four. Space-Time and the Mind 135
    Space-Time and the Impact of the Mind 135
    Space-Time and Chemistry...136
    Space-Time and Biology 138
    Space-Time and Psychology 141
    Space-Time and the Dogmas of the Mind 145

Part Five: Nature

Chapter Five. Space-Time and Nature...149
    Space-Time and the Role of Nature..149
    Space-Time and Micro-Physics 150
    Space-Time and Macro-Physics (Cosmology) 153
    Space-Time and the Contigency of Nature 158

Part Six: Conclusion

Chapter Six. Space-Time and Post-Humans 163
    The Post-Human Challenge...164
    The Post-Human Alteration of Space-Time 165
    The Future of Post-Human Space-Time unto Multiverses.... 171

Bibliography...*179*
Index *185*

Tables

Table 1.1. The Theoretical Debate on Space-Time 18
Table 1.2. No Freedom Without Unfreedom 20
Table 1.3. No Equality Without Inequality................. 23
Table 1.4. The Theory of Floating Consciousness 25
Table 1.5. Pre-Capitalist Value Ideals......................... 27
Table 1.6. Capitalist Value Ideals 29
Table 1.7. Different Versions of Capitalist Value Ideals 31
Table 1.8. Contemporary Alternatives to Capitalist Value Ideals 34
Table 1.9. The Theory of Post-Capitalism I.1: Spiritual/Communal in the Trans-Feminine Calling................. 36
Table 1.10. The Theory of Post-Capitalism I.2: Spiritual/Communal in the Trans-Sinitic Calling 37
Table 1.11. The Theory of Post-Capitalism I.3: Spiritual/Communal in the Trans-Islamic Calling 38
Table 1.12. The Theory of Post-Capitalism I.4: Spiritual/Communal in the Trans-Outerspace Calling 40
Table 1.13. The Theory of Post-Capitalism II: Spiritual/Individualistic in the Post-Human Elitist Calling 42
Table 1.14. Capitalism, Non-Capitalism, and Post-Capitalism 44
Table 1.15. Multiple Causes of the Emergence of Post-Capitalism 48
Table 1.16. The Theory of Post-Democracy I: The Priority of Freedom over Equality........................50
Table 1.17. The Theory of Post-Democracy II: The Priority of Equality over Freedom 52

Table 1.18. The Theory of Post-Democracy III:
    The Transcendence of Freedom and Equality............ 53
Table 1.19. Democracy, Non-Democracy, and Post-Democracy... 55
Table 1.20. Multiple Causes of the Emergence of
    Post-Democracy... 58
Table 1.21. Some Clarifications about Post-Capitalism and
    Post-Democracy... 60
Table 1.22. The Trinity of Pre-Modernity 63
Table 1.23. The Trinity of Modernity 65
Table 1.24. The Trinity of Postmodernity 67
Table 1.25. The Trinity of After-Postmodernity 69
Table 1.26. The Civilizational Project from Pre-Modernity to
    After-Postmodernity 70
Table 1.27. The Structure of Existential Dialectics I:
    The Freedom/Unfreedom and Equality/Inequality
    Dialectics... 72
Table 1.28. The Structure of Existential Dialectics II:
    The Wealth/Poverty Dialectics................................... 74
Table 1.29. The Structure of Existential Dialectics III:
    The Civilization/Barbarity Dialectics 75
Table 1.30. Barbarity, Civilization, and Post-Civilization............ 76
Table 1.31. Five Theses on Post-Civilization 77
Table 1.32. No Freedom Without Unfreedom in the Civilizing
    Processes.. 78
Table 1.33. No Equality Without Inequality in the Civilizing
    Processes..80
Table 1.34. Ontological Constructs in Existential Dialectics 82
Table 1.35. The Logic of Ontology in Existential Dialectics.......... 84
Table 1.36. Civilizational Holism 86
Table 1.37. Theories on Civilizational Holism 89
Table 2.1. Space-Time and Culture 108
Table 3.1. Space-Time and Society 132
Table 4.1. Space-Time and the Mind 146
Table 5.1. Space-Time and Nature 160
Table 6.1. Types of Super Civilization in the Cosmos..................173
Table 6.2. The Technological Frontiers of the Micro-World 175

Table 6.3. Main Reasons for Altering Space-Time......................176
Table 6.4. Physical Challenges to Hyper-Spatial
            Consciousness...177
Table 6.5. Theoretical Speculations of Multiverses. 178

Foreword

Peter Baofu continues to edify the reader with visions of the future, as he has done in the previous eight volumes of his books such as *Beyond Civilization to Post-Civilization* (2006), *Beyond Capitalism to Post-Capitalism* (2005), the 2 volumes *Beyond Democracy to Post-Democracy* (2004), *The Future of Post-Human Consciousness* (2004), *The Future of Capitalism and Democracy* (2002), and the 2 volumes *The Future of Human Civilization* (2000).

Now, in this latest one *The Future of Post-Human Space-Time*, he addresses the issue of space-time—ideas that fascinated Newton and Einstein—but does so from the multiple perspectives of the natural sciences, the humanities and the social sciences. These multiple perspectives in a holistic way have evaded many specialized scholars in the past but have intrigued him, and he questions and explores them here.

In this book Dr. Baofu invites the reader to join him in taking yet another step into the future—to visualize an evolutionary process leading to an outcome as yet undefined. For those not of faint heart, it promises a fascinating trip.

Sylvan von Burg
Georgetown University
Washington, DC
May 2005

Acknowledgments

This book, like many others of mine, does not receive any external funding nor help from the outside. It is solely the result of a self-driven wonder of ideas in the world of knowledge, with the amazing wonder of ideas as its only pleasant reward.

There is one person, however, to whom I want to express special gratitude, namely, Sylvan von Berg at Georgetown University, who is so generous as to write the foreword for this book. His kind support of my intellectual endeavor is therefore deeply appreciated.

In any event, I bear the sole responsibility for all my views expressed in this work.

Abbreviations

BCIV = Peter Baofu. 2006. *Beyond Civilization to Post-Civilization: Conceiving a Better Model of Life Settlement to Supersede Civilization.* NY: The Edwin Mellen Press.

BCPC = Peter Baofu. 2005. *Beyond Capitalism to Post-Capitalism: Conceiving a Better Model of Wealth Acquisition to Supersede Capitalism.* NY: The Edwin Mellen Press.

BDPD = Peter Baofu. 2004. 2 volumes. *Beyond Democracy to Post-Democracy: Conceiving a Better Model of Governance to Supersede Democracy.* NY: The Edwin Mellen Press.

FPHC = Peter Baofu. 2004. *The Future of Post-Human Consciousness.* NY: The Edwin Mellen Press.

FCD = Peter Baofu. 2002. *The Future of Capitalism and Democracy.* MD: The University Press of America.

FHC = Peter Baofu. 2000. 2 volumes. *The Future of Human Civilization.* NY: The Edwin Mellen Press.

PART ONE
Introduction

· CHAPTER ONE ·

Introduction: Space-Time and Humans

Space and time are relative, are reciprocal coordinates, and combine to form the next higher dimension called space-time continuum. They are not constant, absolute, and separate.

—Leonard Shlain (1991: 137)

Space-Time in Classical Mechanics and the Childhood of Science

The word `space-time´ in the title consists of two terms in physics, that is, space and time, to be clarified at the outset in what follows.

The word `space´ derives etymologically from Middle English, which in turn is from Old French *espace* and Latin *spatium* ("area," or "room"). (MWD 2005) By a formal definition, as in Merriam-Webster's collegiate dictionary, it is understood as "a boundless three-dimensional extent in which objects and events occur and have relative position and direction." (MWCD 2003: 1194) The three dimensions, in conventional discourse, of course refer to length, width, and depth.

And the word `time´ has its etymological roots in Middle English, from Old English *tIma* and Old Norse *tImi* (time). (MWD 2005a) Equally by a formal definition, it means "the measured or measurable period during which an action, process, or condition exists or continues." (MWCD 2003a: 1309) So, when combined with the three dimensions of space—the four in question (i.e., time, length, width, and depth) constitute the four dimensions of space-time.

Historically, however, space and time in space-time were treated from an independent, "absolutist" perspective. An excellent example is none other than the idea of absolute space-time by Isaac Newton (1642-1727) in classical mechanics (as summarized in *Table 1.1*).

The absolute conception of space-time in the classical mechanical framework treated every event location as an absolute ordered pair of "spatial location" and "temporal moment," as Lawrence.Sklar (1974: 56-7) aptly put it. This absolute ordered pair analyzes an event in terms of "where the event took place" and "when the event took place." So, for every event, e, it can be expressed as $e = (p, t)$, where p is a spatial location (in space) and t is a temporal moment (in time).

Since time and space were treated as absolute, with their independent existence—the "composition" of space-time can be separated into space and time. (L.Sklar 1974: 56-7) The structure of space, P, therefore, is the set of spatial locations in a three-dimensional Euclidean space, $E3$, whereas the set of temporal moments, T, is simply the one-dimensional real time, $E1$. All events are thus individuals of space-time, that is, the ordered pairs of places and times, with the structure of space-time as $E3 \times E1$, also known as the Cartesian product of space and time in the "absolutist" (or "substantival") view of space-time.

More interestingly, the Newtonian framework of independent space-time allows the existence of space and time even in a world absent of matter. And more, it also allows the existence of places in space and instants in time (that is, event locations) even in a world absent of events. In this sense, the Newtonian framework has a bizarre treatment of space and time as if they were material objects, albeit in a peculiar form, that is, that space "is infinite in extent, Euclidean and three-dimensional in structure" and "persists through time—as do ordinary material objects —but its persistence is characterized by its total unchangingness through time." (L.Sklar 1974: 56-7)

Space and time can be measured in the Newtonian framework, but in an absolute form. For instance, space can be measured in terms of the distance separating different points in space, with their Euclidean geometry. By the same logic, time can also be measured in terms of the temporal separation between the points in time.

While material objects can be used to measure space and time so understood in an absolute form, the Newtonian framework maintains an absolute (unchanging) nature of space-time, in that, even when measuring objects can change their shape in the process (e.g., the expansion of a measuring rod, or the slower rate of ticking in a clock), "absolute" space still retains its unchanging structure, just as "absolute" time still keeps its unchanging magnitude. (L.Sklar 1974: 162) In this way, matter is related to "absolute" space-time in a kind of "container" or "arena," metaphorically speaking, in that objects are "in" space and events occur "in" time.

Space-Time in the Theory of Relativity and the Adolescence of Science

In the early 20th century, Albert Einstein challenged the Newtonian absolutist notion of space-time with his "relativist" version, that is, *the theory of relativity*, with its special version in 1905 (for combining space and time into space-time and matter and energy in matter-energy) and its general version in 1915 (for combining both space-time and matter-energy into a grand union).

In the "special" version of the theory of relativity (1905), matter and energy are shown of its equivalence, that is, as interchangeable, with the famous equation, $E = mc^2$. That is, just as matter can be converted into energy (with one of its most powerful forms, that is, the explosion of a nuclear bomb), its reverse is also possible (e.g., when matter can emerge out of energy in the nothingness of the void, as was the formation of elemental particles in the beginning of the Big Bang). (L.Shlain 1991: 325)

And in space-time, space and time are also relative (and interchangeable); for instance, as time slows down at the speed of light, space changes its shape: "Time is shortened [when clocks slow down at the speed of light], but height and width remain the same....[However, length is] `spaghettified´ at the speed of light." (J.Gribbin 1994: 35, 46) In a well-known thought experiment, "things seen off to the side from the train traveling at one half the speed of light appear vertically elongated, and at higher speeds their tops begin to curve away from the perpendicular; right angles disappear and are replaced by arcs." (L.Shlain 1991: 126)

Later on, in 1915, Einstein proposed the "general" version of the theory of relativity, in which space-time and matter-energy are unified, in showing the interaction between space-time and matter-energy (with the help of Riemann's tensor field equations): "Space is time equals matter is energy." (L.Shlain 1991: 326-7)

Mass-energy and space-time are no longer independent and absolute, since the general theory of relativity shows precisely "how matter `tells´ spacetime how to curve [in a non-Euclidean geometry] and how curved spacetime `tells´ matter how to behave. The reciprocal relationship between Einstein's two new pairs meant that each informed the other about the characteristics it was to exhibit. This complementary duality, the interplay between spacetime and mass-energy, results in a force we call gravity in our three-dimensional world....Space contracts near mass and dilates away from it. Time dilates near mass and contracts away from it....Clocks positioned farther away from the mass of the earth run faster than clocks closer to the earth." (L.Shlain 1991: 328-330)

However, there is something peculiar in general relativity, in that it does not rule out the possible existence of space-time without certain (though not all) forms of matter-energy at all (in a way which brings reminiscence of the absolute Newtonian space-time, albeit in a different context): "In fact,...the theory is compatible with the possibility of a variety of nonequivalent spacetimes with varying metric structures each being possible spacetimes of a world empty of nongravitational mass-energy." (L.Sklar 1974: 164-5)

That qualified, it is interesting to note that, in introducing the theory of relativity, Einstein once commented on Newton's contribution to physics in the earlier era as "the childhood of science" and thus wrote: "Newton, forgive me."

But can one say the same to Einstein's contribution in the late modern era as *the adolescence of science*, in the absence of better words—albeit more mature than its childhood, in light of his inability to unify the four forces (i.e., weak force, strong force, electromagnetic force, and gravitational force) in the universe for the grand unified theory of physics?

But a more intriguing question can indeed be asked, now that this summary of the intellectual history of space-time in physics from its absolute perspective to the contemporary relativist one is given here: What more can be said about the nature of space-time, when the future of life intelligence like humans and others in heaven and earth are taken into account?

The Perspectival Theory of Space-Time

My original contribution to the theoretical debate on space and time here is not to champion Newton's "absolutist" perspective of space-time or even Einstein's "relativist" counterpart, but to offer my original one, namely, in the absence of a better label, what I prefer to call *the perspectival theory of space-time*, subject to the constraints of what I proposed in my previous works as the ontological logic of "existential dialectics."

My theory contains three theses. Firstly, space and time can be understood from multiple perspectives, in relation to culture, society, nature, and the mind, with each perspective revealing something about the nature of space-time and simultaneously delimiting its view. This is subject to "the regression-progression principle" in existential dialectics. Secondly, each perspective of space and time exists in culture, society, nature, and the mind with good reasons, and some are more successful and hegemonic (dominant) than others. This is subject to "the symmetry-asymmetry principle" in existential dialectics.

And finally, space and time will not last, to be eventually superseded (altered) by post-humans in different forms (e.g., stretching/shrinking space-time, engineering more dimensions of space-time, manipulating multiverses, or doing something else that our world in this relatively technologically primitive era has never known), both in this universe and in multiverses. Thus, even the physical existence of space-time cannot last forever, with ever more transformations in the process. This is subject to "the change-constancy principle" in existential dialectics.

The conventional wisdom (especially by physicists) of treating the physical perspective of space and time as the foundation of all other perspectives and of regarding them as less important is a form of reductionism, committing what I call *the foundation fallacy*, in misleadingly dismissing the multiple perspectives of space-time. With these three arguments in mind—the word "perspectival" in the title of my theory is therefore revealing indeed.

The Nature of Existential Dialectics

Thus, a good point of departure to understand the perspectival theory of space-time is to summarize my previous works on the ontological logic of "existential dialectics."

The ontological logic of existential dialectics is based on the theoretical foundation of my earliest books, such as the 2-volume work titled *The Future of Human Civilization* in 2000 (hereafter abbreviated as *FHC*), *The Future of Capitalism and Democracy* in 2002 (hereafter abbreviated as *FCD*) and also *The Future of Post-Human Consciousness* in 2004 (hereafter abbreviated as *FPHC*)—as will be summarized below.

These earliest books led to the further elaboration of existential dialectics in my later works such as the 2-volume work titled *Beyond Democracy to Post-Democracy: Conceiving a Better Model of Governance to Supersede Democracy* in 2004 (hereafter abbreviated as *BDPD*) and another book of mine titled *Beyond Capitalism to Post-Capitalism: Conceiving a Better Model of*

Wealth Acquisition to Supersede Capitalism in 2005 (hereafter abbreviated as *BCPC*).

To start, the term `existential´ in "existential dialectics" has in mind the existence of intelligent life (both primitive and advanced) in general (as the word literally suggests). It should not be confused with the Existentialist movement (or Existentialism), which was already rejected in *FHC, FCD,* and also *FPHC*.

This qualification aside—existential dialectics is in fact a broad summary of the freedom/unfreedom and equality/inequality dialectics which was first proposed in *FHC* but later further developed in *FCD* and also *FPHC*.

The freedom/unfreedom dialectics shows that, all things considered, there is no freedom without unfreedom (as shown in *Table 1.2*) and no equality without inequality (as shown in *Table 1.3*). The seven dimensions of life existence as introduced in *FHC* (i.e., the technological, the everyday, the true, the holy, the sublime/beautiful, the good, and the just), are subject to these existential constraints, be the history about pre-modernity, modernity, postmodernity, and, soon, what I originally proposed in *FHC* as "after-postmodernity."

A theoretical implication is disturbing enough for many of our contemporaries, in that there is no civilization without barbarity (as proposed in *BDPD* and further elaborated in another book of mine titled *Beyond Civilization to Post-Civilization: Conceiving a Better Model of Life Settlement to Supersede Civilization* in 2006, to be hereafter abbreviated as *BCIV*).

This is especially so in the "post-human" age, long after the extinction of the human species, to be replaced in future history by post-humans of various forms (e.g., thinking robots, thinking machines, cyborgs, genetically altered superior beings, floating consciousness, and hyper-spatial consciousness). This post-human vision of mine was first originally analyzed in *FHC* and revisited (further elaborated) in both *FCD* and *FPHC*.

In this light, if one asks the fundamental question, What is the future of human civilization?—my answer in *FCD* (89) is thus: "As addressed in Ch.7 of *FHC*, a later epoch of the age of after-postmodernity (that is, at some point further away from after-postmodernity) will begin, as what I called the `post-human´ history (with the term `post-human´ originally used in my doctoral dissertation at M.I.T., which was finished in November 1995, under the title *After Postmodernity*, still available at M.I.T. library, and was later revised and published as *FHC*). The post-human history will be such that humans are nothing in the end, other than what culture, society, and nature (with some luck) have shaped them into, to be eventually superseded by post-humans (e.g., cyborgs, thinking machines, genetically altered superior beings, and others), if humans are not destroyed long before then."

And "[t]he post-human history," so I continued, "will therefore mark the end of human history as we know it and, for that matter, the end of human dominance and, practically speaking, the end of humans as well. The entire history of human civilization, from its beginning to the end, can be summarized by four words, linked by three arrows (as already discussed in *FHC*)":

Pre-Modernity → Modernity → Postmodernity → After-Postmodernity

---

In *BDPD*, this thesis of mine was rightly called the theory of the evolution from pre-modernity to after-postmodernity, at the historical level.

And "[t]he end of humanity in the coming human extinction is the beginning of post-humanity. To say an untimely farewell to humanity is to foretell the future welcome of post-humanity." (P.Baofu 2002: 89) This thesis of mine was designated in *BDPD* as the theory of post-humanity, at the systemic level.

In Ch.9 of *FCD* (367-8), I further suggested "that civilizational history will continue into the following cyclical progression of expansion, before it is to be superseded (solely as a high probability, since humans might be destroyed sooner either by themselves or in a gigantic natural calamity) by posthumans at some distant point in after-postmodernity (which I already discussed in *FHC*)" unto multiverses (different constellations of universes):

Local → Regional → Global → Solar → Galactic → Clustery... → Multiversal

---

In *BDPD*, this thesis of mine was labeled as the theory of the cyclical progression of system integration and fragmentation, at the systemic level.

In Ch.10 of *FCD*, I predictively proposed different forms of "post-capitalism" and "post-democracy" as the future historical constructs to replace capitalism and democracy unto the post-human age, with "floating consciousness" and "hyper-spatial consciousness" (as elaborated in *FPHC*) as a climax of evolution in consciousness, especially after the eventual extinction of human consciousness:

———

Primordial consciousness → Human consciousness →
Post-human consciousness (with floating consciousness and
hyper-spatial consciousness as a climax in the evolution
of consciousness)

———

In *BDPD*, these theses of mine were known as the theory of floating consciousness (as summarized in *Table 1.4*) and the theory of hyper-spatial consciousness, both at the cosmological and psychological levels.

At the institutional level, the theses above on post-capitalism and post-democracy were called (in, say, *BDPD* and *BCPC*) the theory of post-capitalism (as summarized in *Table 1.9*, *Table 1.10*, *Table 1.11*, *Table 1.12*, and *Table 1.13*), with *Table 1.5*, *Table 1.6*, *Table 1.7*, and *Table 1.8* on a comprehensive comparative analysis of capitalism with other forms hitherto existing in history, with *Table 1.14* on the distinctions among capitalism, non-capitalism, and post-capitalism, as well as *Table 1.15* on multiple causes of the emergence of post-capitalism—and, for that matter, the theory of post-democracy (as summarized in *Table 1.16*, *Table 1.17*, and *Table 1.18*), with *Table 1.19* on the distinctions among democracy, non-democracy, and post-democracy, as well as *Table 1.20* on multiple causes of the emergence of post-democracy and *Table 1.21* on some clarifications in regard to post-capitalism and post-democracy.

In addition, both *FCD* and *FPHC* provided a more elaborated analysis of the structure of post-human civilization in terms of the trinity of after-postmodernity (i.e., "free-spirited after-postmodernity," "post-capitalist after-postmodernity," and "hegemonic after-postmodernity").

Both conceptually and theoretically, the trinity of after-postmodernity constitutes a sequential extension of the trinity of modernity (i.e., "free-spirited modernity," "capitalist modernity,"and "hegemonic modernity") and the trinity of postmodernity (i.e., "free-spirited postmodernity," "capitalist postmodernity," and "hegemonic postmodernity") as first proposed in *FHC*, with the trinity of pre-modernity (i.e., "pre-free-spirited pre-modernity," "pre-capitalist pre-modernity" and "hegemonic pre-modernity") later elaborated in *BCIV*.

In *BDPD*, this thesis about the trinity of pre-modernity, modernity, postmodernity, and after-postmodernity was collectively labeled as the theory of the trinity of modernity to its after-postmodern counterpart, at the cultural level (as summarized in *Table 1.22, Table 1.23, Table 1.24, Table 1.25*, and *Table 1.26*).

At the structural level, no matter which trinity a civilization in question is historically situated in (i.e., be it in pre-modernity, modernity, postmodernity, or after-postmodernity in future times), the freedom/unfreedom and equality/inequality dialectics in the context of "the cyclical progression of hegemony" still applies. Alternatively put, each of the historical epochs has its ever new ways of negotiating for the ever new (different) mixtures of freedom/unfreedom and equality/inequality.

And this is so, not because one is superior (or better) than another in terms of actualizing more freedom and less unfreedom, or more equality with less inequality. Quite on the contrary, each of the historical epochs is subject to the increase of unfreedom in each freedom achieved and the increase of inequality in each equality achieved, albeit in ever new (different) ways. In *BDPD*, this thesis of mine was called the theory of the cyclical progression of hegemony, at the structural level, although it was originally analyzed in *FCD*.

Also in *BDPD*, the nature of existential dialectics was further analyzed in terms of its five main features, in relation to the duality of oppression, namely, (a) that each freedom/equality achieved is also each unfreedom/inequality created, (b) that the subsequent oppressiveness is dualistic, both by the Same against the Others and itself and by the Others against the Same and themselves, (c) that both oppression and self-oppression can be achieved by way of downgrading differences (between the Same and the Others) and of accentuating them, (d) that the relationships are relatively asymmetric among them but relatively symmetric within them, even when the Same can be relatively asymmetric towards itself in self-oppression, and the Others can be likewise towards themselves, and (e) that symmetry and asymmetry change over time, with ever new players, new causes, and new forms, be the locality here on Earth or in deep space unto multiverses—as summarized in *Table.1.27*.

And the same logic holds in relation to wealth and poverty (as addressed in *BCPC*), as summarized in *Table 1.28* on the wealth/poverty dialectics—and also in relation to civilization and barbarity (as addressed in *BCIV*), as summarized in *Table 1.29*, *Table 1.30*, *Table 1.31*, *Table 1.32*, and *Table 1.33* on the civilization/barbarity dialectics.

In *BDPD*, this thesis on existential dialectics was labeled as the theory of existential dialectics, at the cosmological level. An important point to keep in mind is that, in any situation of existential dialectics, there is a price to pay for whatever that be, so a wise question to ask is of course, Is the price worth paying, overall?

Ultimately, civilization cannot eliminate barbarity but learns to exist with it in ever new ways, and it is no more imperative to preserve civilization than necessary to destroy barbarity, so the ideal of civilization is essentially bankrupt, to be eventually replaced by what I originally proposed as "post-civilization" in *BCIV*.

In *BCPC*, I worked out in more detail the logical structure of existential dialectics at the ontological level and proposed three principles for the ontological logic, namely, (a) the regression-progression principle in relation to the "direction" of history, (b) the symmetry-asymmetry principle in relation to the "relationships" among existents, and (c) the change-constancy principle in relation to the "evolution" of time—as summarized in *Table 1.34* on the ontological constructs of existential dialectics and *Table 1.35* on the logic of ontology in existential dialectics.

All the theses afore-summarized are arranged in a holistic framework, as shown in *Table 1.36* on civilizational holism and *Table 1.37* on my theories about civilizational holism.

With this theoretical background of my previous works on existential dialectics in mind, together with the introduction to the paradigmatic shift of understanding space-time in the first two sections, a sensible inquiry is thus, What more can be said about the nature of space-time, especially (though not necessarily) when post-humans and their future in heaven and earth are considered?

This important inquiry is therefore the focus of this project.

Methodological Holism

This inquiry about post-human space-time is complex enough. A more sensible management of the project requires a breakdown into several levels, so this project resorts to "methodological holism," which I already utilized in all my previous books with their broadness of scope.

My methodological holism requires that any topic of inquiry must be given a comprehensive analysis at all relevant levels encompassing all the domains of human knowledge, be they in the natural sciences, the social sciences, and the humanities.

A good illustration of these levels includes (a1) the microphysical, (a2) the chemical, (a3) the biological, (a4) the psychological, (a5) the organizational, (a6) the institutional, (a7) the structural, (a8) the systemic, (a9) the cultural, (a10) the cosmological, and (a11) other relevant levels which are either a combination of the previous ones or the applied applications of a combination of them.

As repeated time and again in *FCD*, *FPHC*, *BCPC*, and *BCIV*, for instance, the classification here can be reorganized in different ways, only if none of the levels (if relevant to an inquiry in question) is ignored. As an illustration, a refreshing way to reorganize the levels is to examine the future of post-human space-time from the four main perspectives of inquiry, namely, (b1) culture, (b2) society, (b3) the mind, and (b4) nature.

In other words, culture in (b1) refers to culture in (a9). Society in (b2) corresponds to the organizational in (a5), the institutional in (a6), the structural in (a7), and the systemic in (a8). The mind (b3) is associated with the chemical in (a2), the biological in (a3), and the psychological in (a4). And nature in (b4) is linked to the cosmological in (a10)—though it overlaps a bit with the systemic in (a8), the micro-physical in (a1), the chemical in (a2), and the biological in (a3).

Surely, there is also the important factor of luck (or randomness), but it is here already included in each of the four categories in question (that is, culture, society, the mind, and nature).

That clarified, it must also be stressed that the comparison in the classification is not absolute, but relative. So, some of the issues in each perspective can be reclassified somewhere else, since they are not strictly mutually exclusive.

Chapter Outline

With this methodological holism in mind—the project is organized into four main parts, corresponding to (b1) space-time and culture, (b2) space-time and society, (b3) space-time and the mind, and (b4) space-time and nature, together with an introduction at the beginning and a conclusion in the end.

In other words, the inquiry about the future of post-human space-time is organized in relation to culture, society, nature, and the mind.

The book thus has six chapters in total, starting with the introductory chapter here, that is, Chapter One titled *Introduction: Space-Time and Humans*, in which the concept of space-time is analyzed, together with an introduction to the historical debate and my own theory on space-time, in relation to my previous works on methodological holism and existential dialectics.

Chapter Two, titled *Space-Time and Culture*, examines the multiple perspectives of space and time from the dimension of culture over the ages.

Chapter Three, titled *Space-Time and Society*, then looks into the very understanding of space and time from the various dimensions of society, be they about social organizations, social institutions, social structure, and social systems.

Chapter Four, titled *Space-Time and the Mind*, proceeds to incorporate the analysis of the nature of space and time from the different standpoints of chemistry, biology, and psychology.

Chapter Five, titled *Space-Time and Nature*, ends up with the inquiry about space and time as observed from the level of nature, in special relation to micro-physics and macro-physics (cosmology).

The last chapter, titled *Conclusion: Space-Time and Post-Humans*, concludes with the prediction concerning the post-human challenge and the daunting task to alter space-time in future eras for the expansion of intelligent life in the universe, together with the question on the future of post-humans in deep space and beyond unto multiverses.

In the end, there is nothing special about space-time, just as there is no more special place for humans and their post-human successors in the universe and beyond, as all will be superseded in one form or another in different stages of future history, if imagination is stretched to its farthest reach for an educated prediction of a world that has never been known in history hitherto existing.

Previous Works, Book Title, and Neologisms

With the chapter outline in mind—three clarifications are necessary on my previous works, book title, and neologisms.

Firstly, this project is built on the theoretical foundation of my previous books (i.e., *FHC*, *FCD*, *FPHC*, *BDPD*, *BCPC*, and *BCIV*), as already summarized in *Sec.1.4*.

In other words, this project is written in conversation with them. While a summary of them is given throughout the project, whenever needed—it is my expectation that the reader is to read these previous works of mine for more analysis.

Secondly, lest any hazy misunderstanding occurs, it should be emphasized that post-human space-time is not better for intelligent life in the sense of being superior to other ones as having been tried before on earth, but in the meaning of historical contingency as being a better fit in relation to different needs of culture, society, nature, and the mind in different spacetimes unto the post-human age here on earth and beyond in multiverses.

As an analogy, capitalism is a better fit in relation to the different historical needs of culture, society, nature, and the mind in this post-Cold War age of ours, just as feudalism enjoyed its better fit in the medieval era.

The term `better´ in the title is therefore a historically relative concept (as this is something I also clarified in *FHC, FCD, FPHC, BDPD, BCPC,* and *BCIV*). So, when standards are used to judge different historical eras, one must understand, with a healthy dose of sensitivity, how standards are so often embedded within the ideology and power interests of an era. And the historical sensitivity on an issue in one era differs from that of another. Yet, in the end, all things considered, all cultures and societies are not immune from the constraints of existential dialectics. There is no utopia in the end; should there be one, dystopia would be embedded within it.

And thirdly, the neologisms used in my works, be they here or elsewhere in my earlier books (e.g., `the perspectical theory of space-time,´ `post-civilization,´ `hyper-spatial consciousness,´ `post-capitalism,´ and the like) are solely for our current intellectual convenience, as they will be renamed differently in different ways in future history.

In *FCD* (508-9), I even wrote that "all these terms `post-capitalism,´ `post-democracy´…and other ones as introduced in this project (e.g.,…`posthuman elitists,´ and `posthuman counter-elitists,´ just to cite a few of them) are more for our current intellectual convenience than to the liking of future humans and post-humans, who will surely invent more tasteful neologisms to call their own eras, entities, and everything else, for that matter. But the didactic point here is to use the terms to foretell what the future might be like, not that its eras and entities must be called so exactly and permanently. After all, William Shakespeare (1995: Act II, Scene II, Line 47) well said long ago: `What is in a name? That which we call a rose by any other name would smell as sweet.´"

In this spirit, each of the neologisms can be understood merely as an "X," to be re-named differently by the powers that be in different eras of future history.

These clarifications aside, let's proceed to Chapter Two on space-time and culture.

Table 1.1. The Theoretical Debate on Space-Time (Part I)

- **Isaac Newton's Absolutist (Substantivist) Theory of Space-Time**
  —space and time are independent from each other. The structure of space-time is $E3 \times E1$ (with the structure of space, P, as the set of spatial locations in a three-dimensional Euclidean space, $E3$, and the structure of time as the set of temporal moments, T, in the one-dimensional real time, $E1$).
  —space and time are also independent from the effects of matter and events. The existence of space and time is possible even in a world absent of matter (and, for that matter, even in a world absent of events), as if they were material objects but with their total unchangingness thorough time.

- **Albert Einstein's Relativist Theory of Space-Time**
  —space and time are interchangeable (not absolute), just as matter and energy are equivalent (not independent) with the famous equation, $E = mc^2$ (as in the special theory of relativity in 1905).
  —space-time and matter-energy are also relative in a grand union (as in the general theory of relativity in 1915). Thus, each pair affects the other pair, as "matter `tells´ spacetime how to curve [in a non-Euclidean geometry] and...curved spacetime `tells´ matter how to behave....Space contracts near mass and dilates away from it. Time dilates near mass and contracts away from it....Clocks positioned farther away from the mass of the earth run faster than clocks closer to the earth." (L.Shlain 1991: 328-330)

(continued on next page)

Table 1.1. The Theoretical Debate on Space-Time (Part II)

- **Peter Baofu's Perspectival Theory of Space-Time**
  - —space and time can be understood from multiple perspectives, be they in relation to culture, society, nature, and the mind, with each perspective revealing something about the nature of space-time and simultaneously delimiting its view. This is subject to "the regression-progression principle" in existential dialectics.
  - —each perspective of space and time exists in society and culture with good reasons, with some being more successful and hegemonic (dominant) than others. This is subject to "the symmetry-asymmetry principle" in existential dialectics.
  - —space and time will not last, to be eventually superseded (altered) by post-humans in different forms (e.g., stretching / shrinking space-time, engineering more dimensions of space-time, and manipulating multiverses), be they here in this universe or in multiverses. Thus, even the physical existence of space-time cannot last forever, with ever more transformations in the process. This is subject to "the change-constancy principle" in existential dialectics.
  - —the conventional wisdom (especially by physicists) of treating the physical perspective of space and time as the foundation of all other perspectives (of space and time) and of regarding them as much less important is a form of reductionism, committing what I call *the foundation fallacy*, in misleadingly dismissing the multiple perspectives of space and time in relation to culture, society, nature, and the mind.

Notes: The examples in each category are solely illustrative (not exhaustive), and the comparison is relative (not absolute), nor are they necessarily mutually exclusive. Some can be easily re-classified elsewhere. As generalities, they allow exceptions.
Source: A summary of *Sec.1.1, Sec.1.2, Sec.1.3*—and for that matter, the rest of the book.

Table 1.2. No Freedom Without Unfreedom (Part I)

- **On Having**
  - *—In Relation to the Technological*
    - (1) if freer from submission to Nature, then less free from ecological degradation (Deep and Social Ecology), even if in a hi-tech form
    - (2) if freer from technological inconvenience/backwardness, then less free from technological control and the loss of privacy
    - (3) if freer from technological (material) backwardness, then less free from the abusive (barbaric) maltreatment of the primitive Others
  - *—In Relation to the Everyday*
    - (1) if freer from abject poverty, then less free from artificial needs/discontents (Frankfurt School)
    - (2) if freer from sensual suppression, then less free from violent sublimation (Freud)
    - (3) if freer from the snobbishness of high culture, then less free from the shabbiness (leveling-off effect) of mass culture (Tocqueville)
    - (4) if freer from the inefficiency of traditional "compassionate economy," then less free from the bondage of a "ruthless [competitive] economy" (Keynes)
    - (5) if freer from anarchy in the state of nature (system fragmentation), then less free from government regulations and controls in system integration

(continued on next page)

Table 1.2. No Freedom Without Unfreedom (Part II)

- **On Belonging**
  - *—In Relation to the Good and the Just*
    - (1) if freer from disciplinary society, then less free from society of control (Foucault)
    - (2) if freer from the tyranny of one or a few, then less free from the tyranny of the majority (or sometimes, minority veto)
    - (3) if freer from elitist decision making, then less free from political gridlock/cleavage
    - (4) if freer from arbitrary (discretionary) administration, then less free from bureaucratic irrationality (Weber) and legal trickery (loopholes)

- **On Being**
  - *—In Relation to the True*
    - (1) if freer from unscientific dogmas, then less free from instrumental abyss (nihilism). Or conversely, if freer from meaninglessness, then less free from dogmas.
    - (2) if freer from the bondage of partiality / partisanship (e.g., prejudice, discrimination), then less free from the danger of impartiality and neutrality (e.g., opportunism, unrealisticness, lack of compassion, inaction)
    - (3) if freer from making generalizations, then less free from being unable to understand much of anything
  - *—In Relation to the Holy*
    - (1) if freer from collective conscience, then less free from social loneliness
    - (2) if freer from religious absoluteness, then less free from spiritual emptiness
  - *—In Relation to the Beautiful/Sublime*
    - (1) if freer from artistic non-autonomy, then less free from aesthetic disillusion (deconstruction)

(continued on next page)

Table 1.2. No Freedom Without Unfreedom (Part III)

Notes: The examples in each category are solely illustrative (not exhaustive), and the comparison is relative (not absolute), nor are they necessarily mutually exclusive. And some can be easily re-classified elsewhere. As generalities, they allow exceptions.

Sources: A reconstruction from Ch.10 of *FCD*, based on *FHC*

Table 1.3. No Equality Without Inequality (Part I)

- **On Having**
  - *—In Relation to the Technological*
    - (1) if more equal in treating Nature with spiritual unity, then less equal in suppressing the dominant drive to transcend it altogether
  - *—In Relation to the Everyday*
    - (1) if more equal in building social plurality, then less equal in leveling-off effects (e.g., the subsequent relative intolerance of high / intellectual ethos in mass culture industry)
    - (2) if more equal in socioeconomic distribution beyond a certain point, then less equal in efficiency (e.g., resentment, the erosion of work ethics)
    - (3) if more equal in urging an affirmative action program, then less equal in creating victim mentality (in oneself), stigma (from others), reverse discrimination (against the once privileged), and mediocracy (against the more able)

- **On Belonging**
  - *—In Relation to the Good and the Just*
    - (1) if more equal in banning monarchic/oligarchic exclusion, then less equal in producing "the tyranny of the majority" or of "minority veto"
    - (2) if more equal in encouraging participatory decision making, then less equal in inducing political divisiveness (gridlock / cleavage in power blocs) and organizational oligarchy
    - (3) if more equal in institutionalizing a decentralized bureaucracy, then less equal in falling into more territorial/turf politics (intrigues)

(continued on next page)

Table 1.3. No Equality Without Inequality (Part II)

- **On Being**
  - *—In Relation to the Beautiful / Sublime*
    - (1) if more equal in accepting diverse styles ("anything goes" mentality), then less equal in artistic good quality (in leveling-off effects against the best)
  - *—In Relation to the True*
    - (1) if more equal in tolerating multiple viewpoints (no matter how extreme), then less equal in epistemic standards
  - *—In Relation to the Holy*
    - (1) if more equal in celebrating any cults and sects (no matter how questionable), then less equal in spiritual depth and authenticity

Notes: The examples in each category are solely illustrative (not exhaustive), and the comparison is relative (not absolute), nor are they mutually exclusive. And some can be easily reclassified elsewhere. As generalities, they allow exceptions.

Sources: A reconstruction from Ch.10 of *FCD*, based on *FHC*

Table 1.4. The Theory of Floating Consciousness (Part I)

- *At the Micro-Physical Level*
  —Ex: intelligent life without the human physical-chemical system

- *At the Chemical Level*
  —Ex: space radiation and toxins

- *At the Bio-Psychological Level*
  —Ex: exo-biological evolution in deep space
  —Ex: genetic engineering of new beings

- *At the Institutional Level*
  —Ex: post-capitalism
  —Ex: post-democracy

- *At the Organizational Level*
  —Ex: less legal-formalistic routines

- *At the Structural Level*
  —Ex: alien forms of violence

- *At the Cultural Level*
  —Ex: transcending freedom
  —Ex: transcending equality

(continued on next page)

Table 1.4. The Theory of Floating Consciousness (Part II)

- *At the Cosmological Level*
  —Ex: parallel universes
  —Ex: pocket universes

- *At the Systemic Level*
  —Ex: space habitats (in zero-gravity)

Notes: Each example draws from the works of different scholars in the field. For instance, at the cosmological level, the idea of parallel universes is from the theoretical speculation in quantum cosmology by Stephen Hawking and others, while the one of pocket universes comes from the theoretical work of Allan Guth at MIT. And at the institutional level, I proposed post-capitalism and post-democracy in *FCD*. In addition, the examples are solely illustrative (not exhaustive), and some of the items can be reclassified somewhere else. Nor are they always mutually exclusive. Since they are generalities, exceptions are expected.
Source: From Ch.1 of *FPHC*

Table 1.5. Pre-Capitalist Value Ideals (Part I)

- **Hunting/Gathering Economics (roughly until 10,000-8,000 B.C.)**
  - *More Spiritual Than Secular*
    - Ex: subsistence level of existence, with little material comfort; highly superstitious
  - *More Communal Than Individualistic*
    - Ex: communal, with little or no social differentiation in tight nomadic groups
  - *More Informal-Legalistic Than Formal-Legalistic*
    - Ex: social relationships on tribal or familial basis, often with no more than 40 people (more or less) in a nomadic group

- **Feudalist Economics (around 12th-15th centuries)**
  - *More Secular Than Spiritual*
    - Ex: the preservation of the feudal monarchy in the web of power relationships among the king (the chief feudal lord), lords, vassals, and serfs
  - *More Individualistic Than Communal*
    - Ex: serfs produce enough for themselves and then pay rent to their feudal superiors, with any surplus left for selling at the market in a nearby town.
  - *More Informal-Legalistic Than Formal-Legalistic*
    - Ex: the particularistic bondage between a lord and a vassal in terms of "homage" (promise to fight for the lord) and "fealty" (promise to remain faithful to the lord), in exchange of "fief" (land) for the vassal

---

(continued on next page)

Table 1.5. Pre-Capitalist Value Ideals
(Part II)

- **Mercantilist Economics (around 177h-18th centuries)**
  —*More Secular Than Spiritual*
    - Ex: the promotion of trade and import of precious metals for the power and wealth of the state
  —*More Individualistic Than Communal*
    - Ex: the driving motive of self-interest in all participants
  —*More Informal-Legalistic Than Formal-Legalistic*
    - Ex: close working relationships among the state and domestic industries against foreign competition; collusion among technocrats, government officials, and merchants

- **Physiocratic Economics (around the 18th century)**
  —*More Secular Than Spiritual*
    - Ex: the promotion of agriculture as the main source of wealth
  —*More Individualistic Than Communal*
    - Ex: the special cultivation of the interest of the land-owner class as inherently linked to that of society
  —*More Informal-Legalistic Than Formal-Legalistic*
    - Ex: collusion between the government and the land-owner class to ensure other economic activities (e.g., manufacturing) to be contingent on a surplus of agricultural production

Notes: The categories and examples are solely illustrative, and the comparison is also relative (not absolute), nor are they mutually exclusive. As generalities, they allow exceptions.

Sources: From *BCPC*, based on a reconstruction from data in F.Pearson (1999: Ch.2), WK (2004), ME (2002), IW (1995), NS (2004), G.Grenier (2002), WK (2004e), and WK (2004f)

Table 1.6. Capitalist Value Ideals
(Part I)

- **More Individualistic Than Communal**
    - *Egoistic*: treating individuals, not as ends in themselves but as means to an end for the self
    - *Competitive*: fighting for market success to the point, in extreme cases, of seeking success for its own sake, instead of focusing on cooperation (collusion)
    - *Insatiable*: always wanting more for the self, accepting no limit of what to acquire

- **More Formal-Legalistic Than Informal-Legalistic**
    - *Diffusive*: knowing myriad others in business in a less specific way, without depth (other than for business)
    - *Emotion-Neutral*: thinking and acting with others on a less affective tone. Business does not mix with fraternization.
    - *Achievement-Oriented*: hiring (or firing) on the basis of merit (or lack of it), not ascription (family relationships)
    - *Unparticularistic*: making business deals on the basis of cost-benefit analysis. Business is not to be polluted with personal intimate relationships.

- **More Secular Than Spiritual**
    - *Pragmatic (Short-Term)*: thinking in terms not of historical veneration but of behavioral consequences in foreseeable time range
    - *Calculative*: guiding action in terms of cost-benefit analysis, instead of moral evaluation
    - *Transformative*: remaking everything at hand into something new, rather than adjusting it to existing norms and virtues

(continued on next page)

Table 1.6. Capitalist Value Ideals
(Part II)

Notes: The categories and examples are solely illustrative since there can be different versions, and the comparison is also relative (not absolute), nor are they mutually exclusive. As generalities, they allow exceptions. And it does not matter whether or not the value ideals are either "market"-capitalistic (more on relative freedom) or "state"-capitalistic (more on relative equality), since they both differ drastically from, say, socialism and, even more radically, communism.

Source: From Ch.10 of *FCD*. Refer to text for more info and references.

Table 1.7. Different Versions of Capitalist Value Ideals (Part I)

- **Classical Economics (e.g., Adam Smith, David Ricardo)**
  —*More Secular Than Spiritual*
  - Ex: economic interest as the driving force of life, with "labor" as the main source of wealth

  —*More Individualistic Than Communal*
  - Ex: "perfect competition" among individuals, with the assurance of the "invisible hand" of the "free market"

  —*More Formal-Legalistic Than Informal-Legalistic*
  - Ex: business is business, based on formal "contractual relations."

- **Neo-Classical Economics (e.g., W. Stanley Jevons, Alfred Marshall)**
  —*More Secular Than Spiritual*
  - Ex: economic interest as the driving force of life, in special relation to the rationality of "utility" and "maximization" (e.g., "profit maximization" of the firm and "utility maximization" of the consumer)

  —*More Individualist Than Communal*
  - Ex: the focus on market "equilibria" as "solutions of individual maximization problems"; the use of "methodological individualism" to explain economic phenomena "by aggregating over the behavior of individuals"

  —*More Formal-Legalistic Than Informal-Legalistic*
  - Ex: business is business, based on formal "contractual relations."

(continued on next page)

Table 1.7. Different Versions of Capitalist Value Ideals (Part II)

___

- **Keynesian Economics (e.g., John Maynard Keynes)**
  —*More Secular Than Spiritual*
    - Ex: economic interest as the driving force of life, with special interest in the problem of "business cycles"
  —*More Individualistic Than Communal*
    - Ex: the role of the free market, but for the interventionist role of the government on occasions of "market failures" (e.g., the Great Depression)
  —*More Formal-Legalistic Than Informal-Legalistic*
    - Ex: business is business, based on formal "contractual relations."

- **Monetarist Economics (e.g., Milton Friedman)**
  —*More Secular Than Spiritual*
    - Ex: economic interest as the driving force of life, with special attention to issues about money supply and inflation
  —*More Individualistic Than Communal*
    - Ex: the role of the free market, with a minimal role of the government (especially the central bank) to solely maintain price stability
  —*More Formal-Legalistic Than Informal-Legalistic*
    - Ex: business is business, based on formal "contractual relations."

___

(continued on next page)

Table 1.7. Different Versions of Capitalist Value Ideals (Part III)

- **New Classical Economics (e.g., John Muth, Robert Lucas)**
    - *—More Secular Than Spiritual*
        - Ex: economic interest as the driving force of life, with particular attention to issues concerning "rational expectations"
    - *—More Individualist Than Communal*
        - Ex: the focus on how individuals engage in expectations of future economic events on the basis of all available info (not just on past data as in adaptive expectations)
    - *—More Formal-Legalistic Than Informal-Legalistic*
        - Ex: business is business, based on formal "contractual relations."

- **Neo-Mercantilism (e.g., Japan and Germany after WWII)**
    - *—More Secular Than Spiritual*
        - Ex: the pursuit of economic and political power of the state, away from the primacy of manufacturing (as in old mercantilism) towards the battle in advanced technology
    - *—More Individualistic Than Communal*
        - Ex: the helping role of the state in economic development, with special favor to the interest of business and technocratic strata
    - *—More Formal-Legalistic Than Informal-Legalistic*
        - Ex: business is still business, yet with some degree of close relationships between the state and the business/technocratic strata (but not to the extreme extent as in old mercantilism).

Notes: The categories and examples are solely illustrative, and the comparison is also relative (not absolute), nor are they mutually exclusive. As generalities, they allow exceptions.

Sources: From *BCPC*, and a reconstruction based on data from WK (2004a), WK (2004b), F.Pearson (1999: Ch.2), WK (2004c), and WK (2004d)

Table 1.8. Contemporary Alternatives to Capitalist Value Ideals (Part I)

- **Marxian Economics (e.g., Karl Marx, Friedrick Engels)**
  —*More Spiritual Than Secular*
    - Ex: the concern with the freedom from labor alienation: "The realm of freedom actually only begins where labor which is determined by necessity and mundane considerations ceases" (K.Marx, *Capital*, v.3)
  —*More Communal Than Individualistic*
    - Ex: the utopian communes, where the state will "wither away"
  —*More Informal-Legalistic Than Formal-Legalistic*
    - Ex: the abolition of the oppressive "contractual relationships" in "capitalist production relations"

- **Eco-Feminist Economics (e.g., Francois d'Eaubonne)**
  —*More Spiritual Than Secular*
    - Ex: the abolition of male oppression of both women and nature, as two main dimensions of the same androcentric violence
  —*More Communal Than Individualistic*
    - Ex: the ecological crisis and the oppression of women as also threats to humanity as a whole
  —*More Informal-Legalistic Than Formal-Legalistic*
    - Ex: the compassion for the Others and the care of nature as vital to humane social relationships, not solely on the basis of formal "contractual relationships"

(continued on next page)

Table 1.8. Contemporary Alternatives to Capitalist Value Ideals (Part II)

- **Islamic Economics**
  - —*More Spiritual Than Secular*
    - Ex: the rationality of *homo Islamicus* (with religious inspirations), not *homo economicus* in capitalist economics
  - —*More Communal Than Individualistic*
    - Ex: the public virtue of payment of the *zakat* (for charity), as one of the five pillars of Islam, and the prohibition of usury (*riba*, meaning: interest)
  - —*More Informal-Legalistic Than Formal-Legalistic*
    - Ex: social relationships based on Islamic norms (say, as indicated in the five pillars)—not on formal capitalist "contractual relationships"

Notes: The categories and examples are solely illustrative, and the comparison is also relative (not absolute), nor are they mutually exclusive. As generalities, they allow exceptions.

Sources: A reconstruction based on data from F.Pearson (1999: Ch.2), K.Marx (1999), and P.Baofu (2000; 2002; 2004)

Table 1.9. The Theory of Post-Capitalism I.1: Spiritual/Communal in the Trans-Feminine Calling

- **More Communal Than Individual**
  - *Sharing*: learning from others, as different ideas mutually enrich
  - *Cooperative*: encouraging a sense of shared leadership and teamwork

- **More Informal-Legalistic Than Formal-Legalistic**
  - *Specific*: listening more from the heart than from the head, to know a person as a concrete, not as an abstract, unit
  - *Affective*: thinking and acting with others on a more affective tone. Business can mix with an emotional touch.
  - *Ascriptive*: hiring (or firing) can be done on the basis of merit (or lack of it), but deep solidarity (sisterhood) is important too.
  - *Particularistic*: making decisions on the basis of cost-benefit analysis, but a given group relationship is vital

- **More Spiritual Than Secular**
  - *Long-Term Looking*: sharing for a long-term relationship (e.g., love, friendship), not just for a short-term gain
  - *Loving/Caring*: showing compassion for the sufferings of others, without quickly blaming and pre-judging
  - *Respectful*: showing acceptance about others' feelings (and thoughts)

Notes: The categories and examples are solely illustrative, since there can be different versions, and the comparison is relative (not absolute), nor are they mutually exclusive. As generalities, they allow exceptions. The specific forms of the trans-feminine version need to be further developed in future after-postmodern history, as they will be different from the ones we now know, since the prefix "trans-" here means going beyond or deconstructing the feminine values, while using them as the inspirational point at the beginning.

Source: From Ch.10 of *FCD*. Refer to text for more info and references

Table 1.10. The Theory of Post-Capitalism I.2: Spiritual/Communal in the Trans-Sinitic Calling

- **More Communal Than Individualistic**
  - *Centralized*: being more top-down in management
  - *Collective*: encouraging more group cooperation
  - *Social*: investing in trust and connection

- **More Informal-Legalistic Than Formal-Legalistic**
  - *Specific*: knowing more of those related or connected
  - *Affective*: behaving in a paternalistic, hierarchical way
  - *Ascriptive*: favoring family members and those related
  - *Particularistic*: building connection (*guanxi*) as imperative

- **More Spiritual Than Secular**
  - *Expansionist*: diffusing civilizational values (e.g., the superiority complex of civilizationalism)
  - *Holistic*: synthesizing things into a panoramic horizon
  - *Historical*: learning from the lessons of the ancient past
  - *Respectful*: deferential to elders and superiors

Notes: The categories and examples are solely illustrative, since there can be different versions, and the comparison is relative (not absolute), nor are they mutually exclusive. As generalities, they allow exceptions. The specific forms of the trans-Sinitic version need to be further developed in future after-postmodern history, as they will be different from the ones we now know, since the prefix "trans-" here means going beyond or deconstructing the Sinitic values, while using them as the inspirational point at the beginning.

Source: From Ch.10 of *FCD*. Refer to text for more info and references.

Table 1.11. The Theory of Post-Capitalism I.3: Spiritual/Communal in the Trans-Islamic Calling (Part I)

- **More Communal Than Individualistic**
  - *Collective:* building the webs of relationships to bind individuals
  - *Sharing:* cultivating the established "wisdom" through common experience
  - *Cooperative:* stressing harmony, solidarity, and commonality

- **More Informal-Legalistic Than Formal-Legalistic**
  - *Specific:* making efforts to know well the participants (family and larger community) in matters of common concern
  - *Affective:* mixing work with language and ritual on explicit religious (Islamic) ideals, texts, stories, and examples
  - *Ascriptive:* privileging local history and custom on relationships among kinship groups
  - *Particularistic:* preferring an unbiased insider with ongoing connections to all parties

- **More Spiritual Than Secular**
  - *Historical:* learning from the lessons of the past as a source of stability and guidance
  - *Deferential:* showing respect for age, experience, status, and leadership in communal affairs
  - *Honorable:* emphasizing face, dignity, prestige, and fairness
  - *Compassionate:* giving mercy and charity ("*Zahah*") to others

(continued on next page)

Table 1.11. The Theory of Post-Capitalism I.3: Spiritual/Communal in the Trans-Islamic Calling (Part II)

Notes: The categories and examples are solely illustrative (not exhaustive), and the comparison is relative (not absolute), nor are they mutually exclusive. As generalities, they allow exceptions. The specific forms of the trans-Islamic version need to be further developed in future after-postmodern history, as they will be different from the ones we now know, since the prefix "trans-" here means going beyond or deconstructing the Islamic values, while using them as the inspirational point at the beginning.

Sources: From Ch.10 of *FCD*. Refer to text for more info and references, especially from the works by George Irani (2000) and C.Murphy (September 19, 2001).

Table 1.12. The Theory of Post-Capitalism I.4: Spiritual/Communal in the Trans-Outerspace Calling (Part I)

- **More Communal Than Individual**
    - *Cooperative*: requiring teamwork in small space habitats
    - *Sharing*: learning from, and enjoying being with, each other in a small group in outer space

- **More Informal-Legalistic Than Formal-Legalistic**
    - *Specific*: knowing more about each other to facilitate living and working together in space, both as fellow astronauts and space-mates
    - *Affective*: being friendly and social to each other as vital to working and living in small space quarters
    - *Ascriptive*: nurturing comaraderie among fellow astronauts as if they are family members over time
    - *Particularistic*: building work relationship with enduring memory in a space mission

- **More Spiritual Than Secular**
    - *Long-Term*: looking beyond selfish materialistic concerns in a precarious space environment with potential life or death
    - *Loving/Caring*: cultivating deep bondage for the success of a long term space mission
    - *Transcendent*: searching for life meaning in outer space

(continued on next page)

Table 1.12. The Theory of Post-Capitalism I.4: Spiritual/Communal in the Trans-Outerspace Calling (Part II)

Notes: The calling and examples in each category are solely illustrative (not exhaustive), since there will be many different outer-space value ideals in the distant future of space colonization. The comparison is also relative (not absolute), nor are they mutually exclusive. As generalities, they allow exceptions. And the specific forms of trans-outer-space calling need to be further developed in future after-postmodern history, as they will be different from the ones we now know, since the prefix "trans-" here means going beyond or deconstructing the current outer-space values, while using them as the inspirational point at the
beginning. The point here is to solely give an extremely rough picture of a small part of the world to come that we still do not know much about.
Source: From Ch.10 of *FCD*. Refer to text for more info and references.

Table 1.13. The Theory of Post-Capitalism II: Spiritual/Individualistic in the Post-Human Elitist Calling (Part I)

- **More Individualistic Than Communal**
  - Setting up rank distinctions among unequals (e.g., between inferior humans and superior post-humans, or later among inferior post-humans and superior ones, relatively speaking)
  - Yearning for being not only distinguished from unequals, but also the first among equals (the best of the very best)
  - Recognizing the constraints of equality/inequality dialectics (or existential dialectics in general)

- **More Spiritual Than Secular**
  - Soul-searching for a high spiritual culture (not the trashy one for the masses). Mass culture is a dirty joke for them.
  - Exploring different spheres of non-human consciousness in the cosmos (something vastly superior than the human one)
  - Recognizing the constraints of freedom / unfreedom dialectics (or existential dialectics in general)

- **Qualifications**
  - Although post-human elitist post-democracy is comparable to post-human elitist post-capitalism in some respects, the former does not necessarily imply the latter (post-human elitist post-capitalism), just as democracy does not have to entail capitalism. They are two different (though related) entities.
  - But up to a certain threshold of incorporating government intervention with advanced info systems in future civilizations for higher spiritual concerns at the expense of the free market and materialist pursuit, the capitalist ideal will be overcome.
  - The overcome will not be Fascist or feudalistic, but post-capitalist, subject to the constraints of existential dialectics.

(continued on next page)

Table 1.13. The Theory of Post-Capitalism II: Spiritual/Individualistic in the Post-Human Elitist Calling (Part II)

Notes: The calling and examples in each category are solely illustrative (not exhaustive). The comparison is also relative (not absolute), nor are they mutually exclusive. As generalities, they allow exceptions. And the specific forms of post-human elitist post-capitalism need to be further developed in future after-postmodern history, as they will be different from the ones we now know, while using them as the inspirational point at the beginning. The point here is to solely give an extremely rough picture of a small part of the world to come that we still do not know much about.

Sources: From Ch.10 of *FCD* (and also *FPHC*, *BDPD*, and *BCPC*). Refer to the text for more info and references.

Table 1.14. Capitalism, Non-Capitalism, and Post-Capitalism (Part I)

- **Capitalism**
  - *Theoretical Constructs*
    - Allocation of scarce resources among alternative wants largely by free market for competition (whose characteristics in its ideal form include, for instance, no barrier to entry or exit, homogeneity, perfect information, a large number of buyers/sellers, and perfect factor mobility)
    - More formal-legalistic than informal-legalistic, more individualistic than communal, and more material (secular) than spiritual
    - Either (1) minimal government or (2) relatively active government
  - *Types*:
    - Only (1): Different versions of market capitalism (e.g., USA)
    - Only (2): Different versions of welfare capitalism (e.g., Sweden)

(continued on next page)

Table 1.14. Capitalism, Non-Capitalism, and Post-Capitalism (Part II)

- **Non-Capitalism**
  - *Theoretical Constructs*
    - Allocation of scarce resources among alternative wants mainly by political authority for policies (which can be regulative, redistributive, symbolic, and participatory)
    - More informal-legalistic than formal-legalistic
    - Either (1') more individualistic (for the elites), often (though not always) for material (secular) concerns, or (2') more communal (for the masses), often (though not always) for spiritual concerns
  - *Types*
    - Only (1'): Different versions on the Right (e.g., Fascist corporate-state economy for the glory of the new Rome, medieval lord-vassal-serf economy for the power of the feudalistic order)
    - Only (2'): Different versions on the Left (e.g., Soviet command economy for the creation of the New Socialist Man)

(continued on next page)

Table 1.14. Capitalism, Non-Capitalism, and Post-Capitalism (Part III)

- **Post-Capitalism**
  - *Theoretical Constructs*
    - Allocation of scarce resources among alternative wants largely by political authority with advanced info systems in future civilizations, subject to existential dialectics. In degree of allocating by authority, post-capitalism is more than capitalism but less than non-capitalism.
    - More spiritual than secular (material)
    - Either (1") more individualistic or (2") more communal
    - Like capitalism and non-capitalism, post-capitalism is also subject to the freedom / unfreedom and equality/inequality dialectics (or existential dialectics in general). There is no utopia, in the end; even were there one, dystopia would exist within it.
    - Unlike capitalism and non-capitalism, post-capitalism makes use of a different degree of political authority with advanced info systems in future civilizations and strives for higher-spiritual cultures (especially in the post-human age), while acknowledging the constraints of existential dialectics and no longer valuing free market (as in capitalism) and economic control (as in non-capitalism) as sacred virtues.
  - *Types*:
    - Only (1"): Different versions of post-human elitist value ideals
    - Only (2"): Different versions of trans-Sinitic value ideals
    - Only (2"): Different versions of trans-feminine value ideals
    - Only (2"): Different versions of trans-Islamic value ideals
    - Only (2"): Different versions of trans-outerspace value ideals

(continued on next page)

Table 1.14. Capitalism, Non-Capitalism, and Post-Capitalism (Part IV)

Notes: The calling and examples in each category are solely illustrative (not exhaustive). The comparison is also relative (not absolute), nor are they mutually exclusive. As generalities, they allow exceptions. And the specific forms of each calling need to be further developed in future after-postmodern history, as they will be different from the ones we now know, while using them as the inspirational point at the beginning. The point here is to solely give an extremely rough picture of a small part of the world to come that we still do not know much about.

Source: From Ch.10 of *FCD*. Refer to the text for more info and references.

Table 1.15. Multiple Causes of the Emergence of Post-Capitalism (Part I)

- *At the Micro-Physical Level*
  - —Ex: intelligent life without the human physical-chemical system
  - —Ex: Mastering of quantum mechanics, electromagnetism, and other fields for the understanding of a broad range of anomalous experiences and the application for artificial intelligence for spiritual quest
  - —Sources: Ch.7 of *FHC*; Chs.9-10 of *FCD*; Ch.1 of *FPHC*

- *At the Chemical Level*
  - —Ex: space radiation and toxins
  - —Sources: Ch.7 of *FHC*; Chs.9-10 of *FCD*

- *At the Bio-Psychological Level*
  - —Ex: exo-biological evolution in deep space
  - —Ex: genetic engineering of new beings
  - —Ex: limits of human cognition
  - —Sources: Ch.2 & Chs.9-10 of *FCD*; Ch.7 of *FHC*

- *At the Institutional Level*
  - —Ex: the flawed logic of the free market
  - —Ex: the need of a post-autistic economics
  - —Source: Ch.10 of *FCD*

- *At the Organizational Level*
  - —Ex: the dark sides of formal-legalistic routines
  - —Sources: Ch.3 of *FHC*; Ch.7 of *FCD*; Ch.3 of *FPHC*

(continued on next page)

Table 1.15. Multiple Causes of the Emergence of Post-Capitalism (Part II)

- *At the Structural Level*
    - —Ex: ever new forms of inequities, at home and abroad
    - —Ex: the emergence of China, women, and Islam as major actors
    - —Sources: Chs.5-6 of *FHC*; Chs.7, 9 & 10 of *FCD*; Chs.4-5 of *BDPD*

- *At the Cultural Level*
    - —Ex: freedom/unfreedom dialectics
    - —Ex: equality/inequality dialectics
    - —Sources: Ch.5 of *FHC*; Chs. 3 & 10 of *FCD*; Ch.4 of *FPHC*; Ch.1 of *BDPD*

- *At the Systemic Level*
    - —Ex: space habitats (in zero-gravity) and colonization
    - —Ex: ultra advanced future info systems
    - —Ex: qualitative demography
    - —Sources: Ch.7 of *FHC*; Chs. 9 & 10 of *FCD*

- *At the Cosmological Level*
    - —Ex: multiverses
    - —Sources: Ch.7 of *FHC*; Chs. 9 & 10 of *FCD*; Ch.4 of *FPHC*

Notes: The examples in each category are solely illustrative (not exhaustive), and some of the items can be reclassified somewhere else. Nor are they always mutually exclusive. Since they are generalities, exceptions are expected.

Sources: From *FHC, FCD, FPHC, BCPC,* and *BDPD*. See also *Table 1.13* and *Table 1.14* on my perspective on civilizational holism.

Table 1.16. The Theory of Post-Democracy I: The Priority of Freedom over Equality (Part I)

- **Differences**
    - *For the aggressive Lions (the strong Elitists)*
        (1) Setting up rank distinctions among unequals (e.g., between inferior humans and superior post-humans, or later among inferior post-humans and superior ones, relatively speaking)
        (2) Yearning for being not only distinguished from unequals, but also the first among equals (the best of the very best)
        (3) Soul-searching for a high spiritual culture (not the trashy one for the masses). Mass culture is a dirty joke for them.
    - *For the manipulative Foxes (the weak Counter-Elitists)*
        (1) Seeking a gentle hegemony by way of more communitarian concerns (for inferior humans and, later, inferior post-humans)
        (2) Being more sympathetic to less formal-legalistic institutions and values

- **Similarities**
    - *For both Lions and Foxes*
        (1) Exploring different spheres of non-human consciousness in the cosmos (something vastly superior than the human one)
        (2) Recognizing the democratic illusions (e.g., no freedom without unfreedom, no equality without inequality, or simply no justice without injustice, and vice versa)

(continued on next page)

Table 1.16. The Theory of Post-Democracy I: The Priority of Freedom over Equality (Part II)

Notes: The two callings and examples in each category are solely illustrative (not exhaustive), since there will be many different post-human value ideals in the distant future of post-human civilization. The comparison is also relative (not absolute) towards post-democracy, so this is not just a version of free-market democracy (nor Fascism/Nazism, as shown in *Table 6.19* on democracy, non-democracy, and post-democracy). Nor are they mutually exclusive. As generalities, they allow exceptions. And the specific forms of post-human post-democratic ideals need to be further developed in future after-postmodern history, as they will be different from the ones we now know. The point here is to solely give an extremely rough picture of a small part of the world to come that we have never known.

Source: From Ch.10 of *FCD*. Refer to text for more info and references.

Table 1.17. The Theory of Post-Democracy II: The Priority of Equality over Freedom

• *Hybrid Versions of*
—Ex: the Trans-Feminine Calling
—Ex: the Trans-Sinitic Calling
—Ex: the Trans-Islamic Calling
—Ex: the Trans-Outerspace Calling

• *Qualifications*
—These four versions of post-capitalist value ideals need not automatically be post-democratic, just as capitalism does not necessarily mean democracy. They are two different entities though closely related.
—But up to a certain threshold of elevating equality at the farther expense of freedom, the democratic ideals will be overcome and cease to exist.
—The overcome will not be socialist or communist, but post-democratic with no freedom without unfreedom and no equality without inequality, subject to the constraints of existential dialectics.

Notes: The callings are solely illustrative (not exhaustive), since there will be many different post-human value ideals in the distant future of post-human lifeforms. The comparison is also relative (not absolute), nor are they mutually exclusive. As generalities, they allow exceptions. And the specific forms of post-human post-democratic ideals need to be further developed in future after-postmodern history, as they will be different from the ones we now know. The point here is to solely give an extremely rough picture of a small part of the world to come that we have never known.
Source: From Ch.10 of *FCD*. Refer to text for more info and references.

Table 1.18. The Theory of Post-Democracy III: The Transcendence of Freedom and Equality (Part I)

- **Transcending Freedom in Floating Existence**
  - *Freedom*: seeking an ultimate elimination of the body. Being without the body. The aim is to transcend freedom in the end into a metaphysical state (i.e., beyond the physique).
  - *Unfreedom*: yet facing difficult trade-offs. The sacrifice of bodily existence and its joyfulness. An eternal boredom in floating existence in dark deep space, though with alternative pleasures. There is no free lunch even in the state of transcending freedom.

- **Transcending Equality in the Rivalry of Cosmic Hegemony**
  - *Inequality*: competing to outlast other lifeforms in floating existence, or just marginalizing them for one's hegemonic expansiveness in the rest of the cosmos (and even beyond). Universalism is only for the mediocre.
  - *Equality*: accepting only those of one's rank as equal partners in the vast spacetime for cosmic supremacy. Even here, the aim is to transcend equality into a metaphysical state.

(continued on next page)

Table 1.18. The Theory of Post-Democracy III: The Transcendence of Freedom and Equality (Part II)

Notes: Do *not* confuse this transcendence of freedom and equality (as one version of post-democracy) with the naïve temptation to transcend the freedom/unfreedom and equality/inequality dialectics. Existential dialectics hold true for freedom and equality in all cultures and societies—past, present, or future (i.e., democracy, non-democracy, and post-democracy), regardless of whether freedom and equality are conventionally understood as "negative" or "positive."

Also, the two features and examples in each are solely illustrative (not exhaustive), since there will be many different post-human value ideals in the distant future of post-human lifeforms. The comparison is also relative (not absolute), nor are they mutually exclusive. As generalities, they allow exceptions. And the specific forms of post-human ideals even for these radically alien floating lifeforms (and others unknown to us) need to be further developed in future after-postmodern history, as they will likely be different from the ones herein illustrated. The point here is to solely give a very rough picture of a small part of the extremely alien world to come that we have never known.

Source: From Ch.10 of *FCD*. Refer to text for more info and references.

Table 1.19. Democracy, Non-Democracy and Post-Democracy (Part I)

- **Democracy**
  - *Theoretical Constructs*
    - The pursuit of freedom and equality (in various degrees), regardless of whether freedom and equality can be understood as "negative" or "positive"
      - (1) more equality than freedom: The relative priority of the good over the right
      - (2) more freedom than equality: The relative priority of the right over the good
  - *Types*
    - Only (1): Different versions of communitarian moral universalism
    - Only (2): Different versions of liberal moral universalism
    - (1) or (2): Different versions of anarchic (non-nation-state) moral universalism
    - (1) or (2): Different versions of postmodern moral localism

- **Non-Democracy**
  - *Theoretical Constructs*
    - The focus on (1') equality or (2') freedom, but not both, regardless of whether freedom and equality can be understood as "negative" or "positive"
  - *Types*
    - Only (1'): Different versions on the Far Left (e.g., Stalinism, Robespierrianism)
    - Only (2'): Different versions on the Far Right (e.g., Nazism, absolute monarchism)

(continued on next page)

Table 1.19. Democracy, Non-Democracy and Post-Democracy (Part II)

- **Post-Democracy**
  - *Theoretical Constructs*
    - The priority of (1") equality over freedom, or (2") freedom over equality, or (3") the transcendence of freedom and equality, regardless of whether freedom and equality are "negative" or "positive." In degree, (1") or (2") is less than (1') or (2') but more than (1) or (2)—respectively.
    - Like democracy and non-democracy, post-democracy is also subject to the freedom / unfreedom and equality/inequality dialectics (or existential dialectics in general). Unlike them, post-democracy acknowledges the constraints of existential dialectics and no longer value freedom and equality as sacred virtues. There is no utopia, in the end; even were there one, dystopia would exist within it.
  - *Types*
    - (1"): Different versions of trans-Sinitic value ideals
    - (1"): Different versions of trans-feminine value ideals
    - (1"): Different versions of trans-Islamic value ideals
    - (1"): Different versions of trans-outerspace value ideals
    - (2"): Different versions of post-human elitist value ideals
    - (3"): Different versions of the value ideals of floating consciousness (etc.)

(continued on next page)

Table 1.19. Democracy, Non-Democracy, and Post-Democracy (Part III)

Notes: The examples are solely illustrative (not exhaustive), nor are they mutually exclusive. As generalities, they allow exceptions. "Negative" freedom is freedom "from" (e.g., freedom from poverty), whereas "positive" freedom is freedom "to" (e.g., freedom to the state of enlightenment). "Negative" equality is "procedural" equality (e.g., equality of opportunity), while "positive" equality is "substantive" equality (e.g., equality of outcome). Existential dialectics impose constraints on freedom and equality in democracy, non-democracy, and post-democracy, regardless of whether freedom and equality can be understood as "negative" or "positive" in conventional discourse. Therefore, do *not* confuse the transcendence of freedom and equality in (3") with the naïve temptation to transcend existential dialectics. There is no utopia, in the end; even should there be one, it would not exist without dystopia embedded within it.

Sources: A summary, based on my previous works, especially Ch.5 of *FHC*, Chs.5-10 of *FCD*, Chs.2-4 of *FPHC*, and Chs.1 & 7 of *BDPD*. The reader should consult the books for more analysis, as this is only a summary here.

Table 1.20. Multiple Causes of the Emergence of Post-Democracy (Part I)

- *At the Micro-Physical Level*
  - —Ex: intelligent life without the human physical-chemical system
  - —Sources: Ch.7 of *FHC*; Chs.9-10 of *FCD*; Ch.1 of *FPHC*

- *At the Chemical Level*
  - —Ex: space radiation and toxins
  - —Sources: Ch.7 of *FHC*; Chs.9-10 of *FCD*

- *At the Bio-Psychological Level*
  - —Ex: exo-biological evolution in deep space
  - —Ex: genetic engineering of new beings
  - —Ex: limits of cognitive partiality
  - —Ex: illusions of emotional neutrality
  - —Ex: human biological inequality
  - —Sources: Ch.2 & Chs.9-10 of *FCD*; Ch.7 of *FHC*; Ch.4 of *BCPC*

- *At the Institutional Level*
  - —Ex: the flawed logic of equality
  - —Ex: the conflicting nature of governance
  - —Sources: Ch.5 of *FHC*; Chs.6 & 10 of *FCD*; Ch.3 of *FPHC*; Chs.2-5 of *BDPD*

- *At the Organizational Level*
  - —Ex: e-civic alienation
  - —Ex: the dark sides of formal-legalistic routines
  - —Sources: Ch.3 of *FHC*; Ch.7 of *FCD*; Ch.3 of *FPHC*

(continued on next page)

Table 1.20. Multiple Causes of the Emergence of Post-Democracy (Part II)

- *At the Structural Level*
    - —Ex: ever new forms of inequities, at home and abroad
    - —Ex: the emergence of China, women, and Islam as major actors
    - —Sources: Chs.5-6 of *FHC*; Chs.7, 9 & 10 of *FCD*; Chs.4-5 of *BDPD*

- *At the Cultural Level*
    - —Ex: freedom/unfreedom dialectics
    - —Ex: equality/inequality dialectics
    - —Ex: system fragmentation and integration
    - —Sources: Ch.5 of *FHC*; Chs.3, 9 &10 of *FCD*; Ch.4 of *FPHC*; Ch.1 of *BDPD*; Ch.4 of *BCPC*

- *At the Systemic Level*
    - —Ex: space habitats (in zero-gravity) and colonization
    - —Ex: ultra advanced future info systems
    - —Ex: qualitative demography
    - —Sources: Ch.7 of *FHC*; Chs.9 &10 of *FCD*

- *At the Cosmological Level*
    - —Ex: the colonization of multiverses
    - —Ex: the expansion of floating consciousness
    - —Ex: the spread of hyper-spatial consciousness
    - —Sources: Ch.7 of *FHC*; Chs.9 &10 of *FCD*; Ch.4 of *FPHC*

Notes: The examples in each category are solely illustrative (not exhaustive), and some of the items can be reclassified somewhere else. Nor are they always mutually exclusive. Since they are generalities, exceptions are expected.

Sources: From *FHC*, *FCD*, *FPHC*, *BCPC*, and *BDPD*. See also *Table 1.13* and *Table 1.14* on my perspective on civilizational holism.

Table 1.21. Some Clarifications about Post-Capitalism (and Post-Democracy) (Part I)

• The prefix `trans-´ in the first category of post-capitalism (with its four versions) refers to something "going beyond" (*not* "uniting" or "combining"). Ex: *Sec.10.3.3* of *FCD*; *Sec.2.4* & *Sec.4.4* of *FPHC*; *Sec.7.2* here.

• Such terms like `post-democracy,´ `post-capitalism,´ `post-human elitist,´ `trans-feminine calling,´ and the like as used in my works are more for our current intellectual convenience than to the liking of future humans and post-humans, who will surely invent more tasteful neologisms to call their own eras, entities, and everything else, for that matter. But the didactic point here is to use the terms to foretell what the future might be like, not that its eras and entities must be called so exactly and permanently. Ex: *Sec.11.1* of *FCD*; *Sec.7.2* here.

• The four versions in the first category of post-capitalist value ideals need *not* automatically be post-democratic, just as capitalism does *not* necessarily mean democracy. They are two different entities—though closely related. But up to a certain threshold of elevating equality at the farther expense of freedom, the democratic ideals will be overcome and cease to exist. The same is true for the post-human elitist calling in the second category of post-capitalism in relation to post-democracy, depending on the extent to which freedom is elevated at the expense of equality. Ex: *Sec.10.4.3.3* of *FCD*; *Table 3.9* of *FPHC*; *Table 7.6* of *BDPD*.

• The comparison in each of the three realms of existence in all forms of post-capitalism is *not* absolute, but relative. Examples include `communal´ vs. `individualistic,´ and the like. Ex: Notes in *Table 10.8, Table 10.9, Table 10.10,* & *Table 10.11* of *FCD*; Chs.2-4 of *FPHC*; *Sec.7.2* here.

(continued on next page)

Table 1.21. Some Clarifications about Post-Capitalism (and Post-Democracy) (Part II)

• The emergence of post-capitalism (and post-democracy, for that matter) has multiple causes (to *not* be reduced to one or only a few). Ex: *Ch.10* of *FCD*, *Chs.2-4* of *FPHC*; *Sec.1.3* & *Sec.7.2* here (or *Table 1.8* & *Table 7.11*).

• The specific forms of post-capitalism (and post-democracy, for that matter) need to be further developed in future after-postmodern history, as they will be different from the ones we now know. The point here is to solely give an extremely rough sketch of a world to come that we have never known. Ex: *Sec.10.3.3* & *Sec.10.4.3.3* of *FCD*; *Table 10.14* & *Table 10.15* of *FCD*; *Sec.7.2* here.

• All forms of post-capitalism are *not* part of a "teleological law," but of "historical trends" only. The same is also true for all forms of post-democracy. Ex: *Sec.7.1* of *FHC*; *Sec.9.5.3.2* & *Sec.10.3.4.2* of *FCD*; *Sec.7.2* here.

• Post-capitalism is *not* better than capitalism in an "absolute" sense but only fits in better, on the basis of the historical contingency of culture, society, nature, and the mind in some future eras. The same is true for post-democracy in relation to democracy. The term `better´ is historically relative. Ex: *Sec.10.3.3* of *FCD*; *Sec.1.7* of *BDPD*; *Sec.1.5* here.

• All forms of post-capitalism (and post-democracy, for that matter) are subject to the constraints of existential dialectics. There is no utopia to be had in the end; even should there be one, dystopia would exist within it. Ex: *Ch.5* of *FHC*; *Sec.10.4.4.2* of *FCD*; *Sec.1.5* of *BDPD*; *Sec.1.3* here.

(continued on next page)

Table 1.21. Some Clarifications about Post-Capitalism (and Post-Democracy) (Part III)

• All forms of post-capitalism, however different from each other though they are, share one common feature, in that they all inspire for a higher spiritual culture. The same is also true for post-democracy. Ex: *Sec.10.3, Sec.10.4 & Sec.10.5* of *FCD*; Chs.2-4 of *FPHC*; *Sec.7.2* here.

• All forms of post-capitalism try to avoid the excess in capitalist consumerism by favoring more basic than artificial needs in *having*, but the quality and quantity of these "basic" needs will be measured by future standards, not by our current ones. Standards are historically relative. Ex: *Sec.10.3, Sec.10.4 & Sec.10.5* of *FCD*; Ch.2 of *FPHC*; *Sec.7.2* here.

• All forms of post-capitalism make use of a different degree of political authority with advanced info systems in future history and strives for higher spiritual cultures (especially in the post-human age), while acknowledging the constraints of existential dialectics and no longer valuing free market (as in capitalism) and economic control (as in non-capitalism) as sacred virtues. Ex: *Sec.10.3.4.2, Sec.10.3, Sec.10.4 & Sec.10.5* of *FCD*; Chs.2-4 of *FPHC*; Sec.1.5 of *BDPD*; *Sec.7.2* here.

Notes: The main points here are solely illustrative (not exhaustive) nor necessarily mutually exclusive, and the comparison is relative (not absolute). As generalities, they allow exceptions. The sections as cited are only illustrative (not exhaustive).
Sources: From *BCPC*. See also *FHC, FCD, FPHC*, and *BDPD*.

Table. 1.22. The Trinity of Pre-Modernity (Part I)

- *Pre-Free-Spirited Pre-Modernity (Pre-Modernism) and Its Internal Split*
    —Competing worldviews and values both within and between linear-centric (e.g., Islamic, Christian, Judaic, Imperial Roman) and cyclical-centric (e.g., Confucian, Taoist, Hindu, and Buddhist) orientations
    —Compare modernism in *Table 1.5* (*BCIV*) with pre-modernism here in relation to the seven dimensions of human existence like the true and the holy (e.g., different versions of epistemic dogmas and religious superstitions), the everyday and the technological (e.g., different versions of non-technophilism and non-consumerism), the beautiful/sublime (e.g., different versions of aesthetic non-autonomy), and the good and the just (e.g., different versions of moral particularism).

- *Pre-Capitalist Pre-Modernity (Pre-Modernization) and Its Own Discontents*
    —Competing versions of societal arrangements (e.g., feudalism, monarchism, and the holy order)

- *Hegemonic Pre-Modernity and Its Countervailing Forces*
    —Different power centers and their enemies (e.g., the Roman Empire and the "barbarian hordes," the "Holy Crusades" and the Muslims, the Middle Kingdom and the invading tribes, different social castes in India, and warring Greek city-states)

(continued on next page)

Table 1.22. The Trinity of Pre-Modernity (Part II)

Notes: The examples in each category are solely illustrative (not exhaustive) nor necessarily mutually exclusive, and the comparison is relative (not absolute). As generalities, they allow exceptions. Also, it does not matter what the "base" era is in the analysis of any trinity. And in the present context, the "base" era is modernity (for instance, with its "free-spirited modernity" and the other two parts). So, for pre-modernity, the trinity takes the form of, say, "pre-free-spirited pre-modernity," together with the other two parts.

Sources: From Ch.2 of *BCIV*. See also the 2 volumes of *FHC*.

Table 1.23. The Trinity of Modernity (Part I)

- **Free-Spirited Modernity (Modernism) and Its Internal Split**
  - *On the True and the Holy*
    - The freedom from the dogmas of the past to the better understanding of, and union with, the world and self (*FHC*: Ch.3)
    - Alternative discourses: about the true (e.g., anti-science discourses) and the holy (non-mainstream theologies) (*FHC*: Ch.3)
  - *On the Technological and the Everyday*
    - The freedom from life harshness to the higher technophilic, consumeristic lifeform (*FHC*: Ch.2)
    - Alternative discourses: about the everyday (e.g., transcendental mindsets) and the technological (e.g., Arcadianism) (*FHC*: Ch.2)
  - *On the Good and the Just*
    - The freedom from the theo-aristocratic tyranny to the moral universality for a just society (*FHC*: Ch.5)
    - Alternative discourses: about the just (e.g., Communism and Anarchism) and the good (e.g., Nazism/Fascism, and Zarathustrianism) (*FHC*: Chs.5-6)
  - *On the Beautiful and the Sublime*
    - The freedom from the external distortion of aesthetic pleasure to the boundless infinity of totality in artistic autonomy (*FHC*: Ch.4)
    - Alternative discourses: about the beautiful/sublime (e.g., kitsch and historical avant-gardism) (*FHC*: Ch.4)

(continued on next page)

Table 1.23. The Trinity of Modernity (Part II)

- **Capitalist Modernity (Modernization) and Its Own Discontents**
  - —*During the Industrial Revolution*
    - Ex: Marx on the institution of inequality (*FHC*: Ch.1)
  - —*During the Modern Rational-Instrumental Epoch*
    - Ex: Weber on the politics of soft liberal institutions (*FHC*: Ch.5)
  - —*During the Great Depression*
    - Ex: Keynes on the myth of the free market (*FHC*: Chs.1,3)
  - —*During the Cold War*
    - Ex: Lasch on the narcissistic culture industry (*FHC*: Chs.2-3)

- **Hegemonic Modernity and Its Countervailing Forces**
  - —*The Legacies of Colonialism and Imperialism*
    - Ex: European colonization of most of the modern world (*FHC*: Ch.1)
  - —*The Struggle for Decolonialization*
    - Ex: The countervailing forces of resentment, rechantment, and regionalism (*FHC*: Chs.1 & 6)

Notes: The examples in each category are solely illustrative (not exhaustive) nor necessarily mutually exclusive, and the comparison is relative (not absolute). As generalities, they allow exceptions.

Sources: From the 2 volumes of *FHC*—and also from *FCD*

Table 1.24. The Trinity of Postmodernity (Part I)

- **Free-Spirited Postmodernity (Postmodernism) and Its Internal Split**
  —*On the True and the Holy*
    - Postmodern performative turn for knowing and its enemies (*FHC*: Ch.3)
    - Postmodern comparative theology and its opponents (*FHC*: Ch.3)
  —*On the Technological and the Everyday*
    - Postmodern corporate technological mindset and its adversaries (*FHC*: Ch.2)
    - Postmodern postmaterialism and its critics (*FHC*: Ch.2)
  —*On the Good and the Just*
    - Postmodern politics of difference and its foes (*FHC*: Ch.5)
  —*On the Beautiful and the Sublime*
    - Postmodern deconstruction and its dissenters (*FHC*: Ch.4)

- **Capitalist Postmodernity (Postmodernization) and Its Own Discontents**:
  —*During the Post-Cold War and Beyond*
    - Ex: post-Fordism and its shortcomings (*FHC*: Ch.6; *FCD*: Chs.6-7)

(continued on next page)

Table 1.24. The Trinity of Postmodernity (Part II)

- **Hegemonic Postmodernity and Its Countervailing Forces**
  —*The Debate on the Global Village*
    - Ex: uni-civilizationalism vs. multi-civilizationalism (*FHC*: Ch.6)
  —*The Resistance Movement*
    - Ex: rechantment and the politics of civilizational claims (e.g., Islamic, Confucian and other ethos in relation to the Same) (*FHC*: Ch.6; *FCD*: Ch.10)
    - Ex: resentment and the politics of resurgence (e.g., the rising Chinese superpower, the growing EU, and other players in relation to the U.S. and her allies) (*FHC*: Ch.6; *FCD*: Ch.8)
    - Ex: regionalism and the politics of inequality (e.g., trans- or inter-national blocs, the North-South divide, NGO's) (*FHC*: Ch.6; *FCD*: Ch.5)

Notes: The examples in each category are solely illustrative (not exhaustive) nor necessarily mutually exclusive, and the comparison is relative (not absolute). As generalities, they allow exceptions.
Sources: From *FCD* and the 2 volumes of *FHC*

Table 1.25. The Trinity of After-Postmodernity

- *Free-Spirited After-Postmodernity (After-Postmodernism) and Its Internal Split*
  —The discourse of naked contingency (*FCD*: Ch.10; *FPHC*: Ch.4)

- *Post-Capitalist After-Postmodernity (After-Postmodernization) and Its Own Discontents*
  —Different versions of post-capitalism and post-democracy, and their enemies (*FCD*: Ch.10; *FPHC*: Chs.3-4)

- *Hegemonic After-Postmodernity and Its Countervailing Forces*
  —The Cyclical Progression of Hegemony in Multiverses (*FCD*: Chs.9-10; *FPHC*: Ch.4)

Notes: The examples in each category are solely illustrative (not exhaustive) nor necessarily mutually exclusive, and the comparison is relative (not absolute). As generalities, they allow exceptions.
Sources: From *FCD* and also *FPHC*

Table 1.26. The Civilizational Project from Pre-Modernity to After-Postmodernity (Part I)

	Pre-Modern	Modern	Post-modern	After-Post-modern
Main narratives	•Sacralness •Courtliness •Vitalism •Animism	•Freedom •Equality •Fraternity	•Multiplicity •Hybridization	•Naked contingency •Cyclical progression of hegemony
Main institutions	•Monarchy •Aristocracy •Feudalism •Holy order •Primitivism	•Capitalism •Liberalism •Socialism •Nazism •Fascism	•Capitalism •Liberalism •Postmodern politics of difference	•Post-Capitalism •Post-Democracy •Others
Main technological & economic revolutions	•Agricultural	•Service •Industrial	•Informational	•Biological •Material •Energy •Space •Others
Main agents	•Males •Upper strata •Mini-states	•Males •Upper strata •Whites •Empires	•Males •Upper strata •Whites •Others •Supra-states •IO's	•Post-humans •Humans •Others

(continued on next page)

Table 1.26. The Civilizational Project from Pre-Modernity to After-Postmodernity (Part II)

	Pre-Modern	*Modern*	*Post-modern*	*After-Postmodern*
Main impacts	•Local	•International	•Global	•Outer-space •Multiverse
Main outcomes	•Towards moderntiy •Rise of linear- & cyclical-centric civilizations	•Towards post-modernity •Dominance of linear-centric civilization	•Towards after-post-moderntiy •Linear-centric civilization in crisis	•Towards human (& maybe post-human) extinction •Rise of post-civilization

Notes: The examples in each category are solely illustrative (not exhaustive) nor necessarily mutually exclusive, and the comparison is relative (not absolute). As generalities, they allow exceptions.

Sources: From Table 10.16 of *FCD* and also from *BCIV* on post-civilization

Table 1.27. The Structure of Existential Dialectics I: The Freedom/Unfreedom and Equality/Inequality Dialectics
(Part I)

• Each freedom and equality produces its own unfreedom and inequality, regardless of whether the pair occurs in political society (with the nation-state), in civil society (with some autonomy from the state), or elsewhere (e.g., in the private sphere of individual homes)—and regardless of whether freedom and equality are understood as "negative" or "positive."

• Oppression is dualistic, as much by the Same against the Others and itself, as by the Others against the Same and themselves.

• Both forms of oppression and self-oppression can be achieved by way of downgrading differences (between the Same and the Others) and of accentuating them.

• The relationships are relatively asymmetric between the Same and the Others and relatively symmetric within them. This is true, even when the Same can be relatively asymmetric towards itself in self-oppression, just as the Others can be likewise towards themselves.

• Symmetry and asymmetry change over time, with ever new players, new causes, and new forms.

(continued on next page)

Table 1.27. The Structure of Existential Dialectics I: The Freedom/Unfreedom and Equality/Inequality Dialectics (Part II)

Notes: The examples in each category are solely illustrative (not exhaustive) nor necessarily mutually exclusive, and the comparison is relative (not absolute). As generalities, they allow exceptions. "Negative" freedom is freedom "from" (e.g., freedom from poverty), whereas "positive" freedom is freedom "to" (e.g., freedom to the state of enlightenment). "Negative" equality is "procedural" equality (e.g., equality of opportunity), while "positive" equality is "substantive" equality (e.g., equality of outcome). Existential dialectics impose constraints on freedom and equality in democracy, non-democracy, and post-democracy. There is no utopia, in the end; even should there be one, dystopia would exist within it.

Sources: From *Table 1.5* of *BDPD*—and also from *FHC*, *FCD*, and *FPHC*

Table 1.28. The Structure of Existential Dialectics II: The Wealth/Poverty Dialectics

• There is no wealth without poverty, just as there is no poverty without wealth.

• The wealth/poverty dialectics occurs in the realms of having, belonging, and being, in relation to the material, relational, and spiritual.

• The wealth/poverty dialectics also expresses itself at the multiple levels of analysis in accordance to methodological holism, be they about the micro-physical, the chemical, the biological, the psychological, the organizational, the institutional, the structural, the systemic, the cultural, and the cosmological.

• The wealth/poverty dialectics is a different manifestation of existential dialectics in general, subject to the principles in its logic of ontology—just as the freedom/unfreedom and equality/inequality dialectics are likewise.

• There is no economic utopia, in the end; even should there be one, dystopia would exist within it.

―――――
Notes: The main points here are solely illustrative (not exhaustive) nor necessarily mutually exclusive, and the comparison is relative (not absolute). As generalities, they allow exceptions.
Sources: From *BCPC*. See also *FCD* and *FHC*.

Table 1.29. The Structure of Existential Dialectics III: The Civilization/Barbarity Dialectics

• There is no civilization without barbarity.

• The civilization/barbarity dialectics applies in the four civilizing processes (e.g., the rationalizing process, the pacifying process, the stewardizing process, and the subliming process).

• The civilization/barbarity dialectics is another (different) manifestation of existential dialectics in general, subject to the principles in its logic of ontology—just as the freedom/unfreedom and equality/inequality dialectics and the wealth/poverty dialectics are likewise.

• There is no utopia, in the end; even should there be one, dystopia would exist within it.

Notes: The main points here are solely illustrative (not exhaustive) nor necessarily mutually exclusive, and the comparison is relative (not absolute). As generalities, they allow exceptions.
Sources: From *BCIV*. See also *FCD*, *FHC*, and *BDPD*.

Table 1.30. Barbarity, Civilization, and Post-Civilization

- **The Rationalizing Process (at the Level of Culture)**
  —*Barbarity*
    - More mythicizing than rationalizing, relatively speaking
  —*Civilization*
    - More rationalizing than mythicizing, relatively speaking
  —*Post-Civilization*
    - Beyond the dichotomy, subject to existential dialectics

- **The Pacifying Process (at the Level of Society)**
  —*Barbarity*
    - More pillaging than pacifying, relatively speaking
  —*Civilization*
    - More pacifying than pillaging, relatively speaking
  —*Post-Civilization*
    - Beyond the dichotomy, subject to existential dialectics

- **The Stewardizing Process (at the Level of Nature)**
  —*Barbarity*
    - More revering than stewardizing, relatively speaking
  —*Civilization*
    - More stewardizing than revering, relatively speaking
  —*Post-Civilization*
    - Beyond the dichotomy, subject to existential dialectics

- **The Subliming Process (at the Level of the Mind)**
  —*Barbarity*
    - More impulsing than subliming, relatively speaking
  —*Civilization*
    - More subliming than impulsing, relatively speaking
  —*Post-Civilization*
    - Beyond the dichotomy, subject to existential dialectics

Notes: The comparison in each category is relative (not absolute), nor are they necessarily mutually exclusive. And some can be easily re-classified elsewhere. As generalities, they allow exceptions.
Sources: From *BCIV*. See also *FHC*, *FCD*, *FPHC*, *BDPD*, and *BCPC*.

Table 1.31. Five Theses on Post-Civilization

• Post-civilization no longer treats civilization as good and barbarity as evil (relatively speaking), nor does it nostalgically regard barbarity as good and civilization as evil (relatively speaking again). Civilization is as evil and good as barbarity.

• Post-civilization also no longer accepts the dichotomy between civilization and barbarity. Civilization cannot exist without barbarity. It is no longer necessary to preserve civilization, any more than it is imperative to destroy barbarity.

• Post-civilization is also subject to the constraints of existential dialectics. There is no freedom without unfreedom, and no equality without inequality, for instance. There will be no utopia; even should there be one, there would be dystopia embedded within it.

• Post-civilization will eventually replace civilization (as a form of life settlement), to be dominated by post-capitalist and post-democratic lifeforms here on earth and in deep space (besides other alien lifeforms that we have never known), unto the post-human age in multiverses. Those few post-humans who keep civilization will live in a "post-human civilization," while the rest (the majority), who choose post-civilization, will evolve towards the state of "post-human post-civilization." One therefore should *not* confuse "post-human civilization" with "post-human post-civilization," as the two are not the same.

• Post-civilization will confront psychosis as a primary problem in the culture of virtuality unto the post-human age, just as civilization has neurosis as a primary one of its own (although both neurosis and psychosis are major problems in both).

Notes: The comparison in each category is relative (not absolute), nor are they necessarily mutually exclusive. And some can be easily re-classified elsewhere. As generalities, they allow exceptions.
Sources: From *BCIV*. See also *FHC*, *FCD*, *FPHC*, *BDPD*, and *BCPC*.

Table 1.32. No Freedom Without Unfreedom in the Civilizing Processes (Part I)

- *The Rationalizing Process (at the Level of Culture)*
  —if freer from the dominance of unreason (as in barbarism) in the civilizing process, then less free from the rationalizing process (be it in the form of the principle of either transcendence or immanence)
  —if freer from the principle of immanence in the rationalizing process, then less free from the inclination to commit terror in the name of reason and the relative underdevelopment of non-reason (e.g., in relation to yoga and meditation)
  —if freer from the principle of transcendence in the rationalizing process, then less free from the relative underdevelopment of reason (e.g., in relation to systematic methodology) and the occurrence of oppression in the name of non-reason

- *The Pacifying Process (at the Level of Society)*
  —if freer from the dominance of pillage (as in savagery) in the civilizing process, then less free from the pacifying process (be it in the form of external control or self-control)
  —if freer from self-control in the pacifying process, then less free from the temptation of expansionist oppression and rebellious mindset in external control
  —if freer from external control in the pacifying process, then less free from the gruesome psychological self-torture and conformism in self-control

(continued on next page)

Table 1.32. No Freedom Without Unfreedom in the Civilizing Processes (Part II)

- *The Stewardizing Process (at the Level of Nature)*
  —if freer from the dominance of nature (as in the state of nature) in the civilizing process, then less free from the stewardizing process (be it in the form of the stewardship of creation or the covenant with nature)
  —if freer from the stewardship of creation in the stewardizing process, then less free from material underdevelopment, relatively speaking, and spiritual exclusion in the covenant with nature
  —if freer from the covenant with nature in the stewardizing process, then less free from ecological degradation and spiritual disconnection from nature in the stewardship of creation

- *The Subliming Process (at the Level of the Mind)*
  —if freer from the dominance of spontaneity (as in the wild state of the mind) in the civilizing process, then less free from the subliming process, be it in the form of (cyclical-centric) self-refinement or (linear-centric) self-discipline
  —if freer from (cyclical-centric) self-refinement in the subliming process, then less free from the (linear-centric) self-regimen (as a form of neurosis)
  —if freer from (linear-centric) self-discipline in the subliming process, then less free from the (cyclical-centric) self-torture (equally as a form of neurosis)

Notes: The examples in each category are solely illustrative (not exhaustive), and the comparison is relative (not absolute), nor are they necessarily mutually exclusive. And some can be easily re-classified elsewhere. As generalities, they allow exceptions.
Sources: From *BCIV*. See also *FHC*, *FCD*, *FPHC*, *BDPD*, and *BCPC*.

Table 1.33. No Equality Without Inequality in the Civilizing Processes (Part I)

- *The Rationalizing Process (at the Level of Culture)*
  —if more equal for the role of rationalization in the rationalizing process (of civilizational making), then less equal for that of mythicization (as in barbarism)
  —if more equal for the principle of transcendence in (linear-centric) rationalizing process, then less equal for the principle of immanence
  —if more equal for the principle of immanence in (cyclical-centric) rationalizing process, then less equal for the principle of transcendence

- *The Pacifying Process (at the Level of Society)*
  —if more equal for pacification in civilizational making, then less equal for the institution of pillaging and others (as in savagery)
  —if more equal for external control, relatively speaking, in pacifying process, then less equal for self-control
  —if more equal for self-control, relatively speaking, in pacifying process, then less equal for external-control

- *The Stewardizing Process (at the Level of Nature)*
  —if more equal for stewardship in the stewardizing process (of civilizational making), then less equal for reverent (submissive) existence (as in barbarism)
  —if more equal for the stewardship of creation in (linear-centric) stewardizing process, then less equal for the (cyclical-centric) covenant with nature for harmonious co-existence
  —if more equal for the (cyclical-centric) covenant with nature in the stewardizing process, then less equal for the (linear-centric) stewardship of nature for domination

(continued on next page)

Table 1.33. No Equality Without Inequality in the Civilizing Processes (Part II)

- *The Subliming Process (at the Level of the Mind)*
  —if more equal for the role of reason in the subliming process, then less equal for that of unreason (as in the natural state of wildness)
  —if more equal for the primacy of reason in (linear-centric) subliming process, then less equal for other faculties (e.g., intuition, existential feelings, and analogous thinking) in cyclical-centric one
  —if more equal for the exercise of other faculties (e.g., intuition, existential feelings, and analogous thinking) in cyclical-centric subliming process, then less equal for the role of reason in linear-centric counterpart

Notes: The examples in each category are solely illustrative (not exhaustive), and the comparison is relative (not absolute), nor are they mutually exclusive. And some can be easily reclassified elsewhere. As generalities, they allow exceptions.
Sources: From *BCIV*. See also *FHC*, *FCD*, *FPHC*, *BDPD*, and *BCPC*.

Table 1.34. Ontological Constructs in Existential Dialectics
(Part I)

- **On the "Direction" of History**
  - *The Regression-Progression Principle*
    - neither the cyclical nor the linear views are adequate for explaining many phenomena at all levels.
    - history progresses to more advanced forms, but with a regressive touch. Examples include no freedom without unfreedom, no equality without inequality, and no civilization without barbarity.
    - this is not an inevitable law, but merely a highly likely empirical trend.

- **On the "Relationships" among Existents**
  - *The Symmetry-Asymmetry Principle*
    - the relationships are relatively asymmetric between the Same and the Others but relatively symmetric within the Same (or the Others). There is no asymmetry without symmetry. This is true, even when the Same can be relatively asymmetric towards itself in self-oppression, just as the Others can be likewise towards themselves.
    - the subsequent oppressiveness is dualistic, as much by the Same against the Others and itself, as by the Others against the Same and themselves.
    - both oppression and self-oppression can be achieved by way of downgrading differences between the Same and the Others and of accentuating them.

- **On the "Evolution" of Time**
  - *The Change-Constancy Principle*
    - asymmetry undergoes changes over time, so does symmetry.
    - old players fade away, and new ones emerges, with ever new causes and ever new forms.

(continued on next page)

Table 1.34. Ontological Constructs in Existential Dialectics (Part II)

Notes: The categories and examples in each are solely illustrative (not exhaustive). The comparison is also relative (not absolute), nor are they mutually exclusive. As generalities, they allow exceptions.

Sources: From Ch.6 of *BCPC* and also from *FHC*, *FCD*, *FPHC*, and *BDPD*

Table 1.35. The Logic of Ontology in Existential Dialectics (Part I)

- **Sets and Elements**
  - *Sets*
    - Ex: the Same
    - Ex: the Others
  - *Elements*
    - Ex: whites in 20th century America (in the set of "the Same")
    - Ex: Iraq during the U.S. invasion in 2003 (in the set of "the Others")

- **Relations, Operations, Functions**
  - *Relations (e.g., "belongs," "equals to," "is greater than")*
    - Ex: symmetric interactions within the Same (or the Others)
    - Ex: asymmetric interactions between the Same and the Others
  - *Operations (e.g., "and," "or," "not," "if...then")*
    - Ex: if the Same oppresses the Others, it will also oppress itself.
    - Ex: the Same is not the Others.
  - *Functions (e.g., goals)*
    - Ex: the Same is hegemonic in relation to the Others.

- **Truth Values**
  - *"1" if True*
    - Ex: the proposition that imperial Japan was hegemonic to China during WWII
  - *"0" if False*
    - Ex: the proposition that Grenada invaded France in 2003

(continued on next page)

Table 1.35. The Logic of Ontology in Existential Dialectics (Part II)

- **Axioms, Postulates, Theorems, Principles**
  —*Axioms*
    - Ex: the reflexive axiom—"any quantity is equal to itself"
  —*Postulates*
    - Ex: the SSS postulate—"if the three sides of a triangle are congruent to their corresponding parts, then the triangles are congruent
  —*Theorems (and Principles)*
    - Ex: the regression-progression principle
    - Ex: the symmetry-asymmetry principle
    - Ex: the constancy-change principle

Notes: The categories and examples in each are solely illustrative (not exhaustive). The comparison is also relative (not absolute), nor are they mutually exclusive. As generalities, they allow exceptions.

Sources: From Ch.6 of *BCPC* and also from *FHC*, *FCD*, *FPHC*, and *BDPD*

Table 1.36. Civilizational Holism
(Part I)

- *At the Micro-Physical Theoretical Level*
  —Ex: Mastering of quantum mechanics, electromagnetism, and other fields for the understanding of a broad range of anomalous experiences and the application for artificial intelligence (*Sec.1.4.1* of *FPHC*)

- *At the Chemical Theoretical Level*
  —Ex: Unprecedented expansion of (and violence to) the mind through ever new forms of drugs (and virtual technologies, for that matter) (Ch.9 of *FCD*)

- *At the Biological Theoretical Level*
  —Ex: Humans are not biologically equal, on the basis of race, gender, ethnicity, age, and whatnot. (*Sec.2.6* & Ch.10 of *FCD*) And post-humans will experience the same fate, in an even more amazing way.

- *At the Psychological Theoretical Level*
  —Ex: Human cognitive impartiality and emotional neutrality are quite limited. (*Secs.2.4-2.5* of *FCD*)

- *At the Organizational Theoretical Level*
  —Ex: Administrative colonization of deep space, with less legal-formalism in some corners. (Chs.9-10 of *FCD*)

- *At the Institutional Theoretical Level*
  —Ex: Both capitalism and democracy will not last, to be superseded by different versions of post-capitalism and post-democracy in after-postmodernity. (Ch.10 of *FCD*)

(continued on next page)

Table 1.36. Civilizational Holism (Part II)

- *At the Structural Theoretical Level*
  - Ex: Social stratification reappears in ever new forms, also with new causes and new players in the cyclical progression of hegemony. (Chs.8-10 of *FCD*)

- *At the Systemic Theoretical Level*
  - Ex: Outerspace expansion: local → regional → global → solar → galactic → clustery...→ multiversal (Ch.9 of *FCD*)
  - Ex: Demographic transition: human extinction, and the rise of post-humans (e.g., cyborgs, thinking machines, thinking robots, genetically altered superior beings, floating consciousness, hyper-spatial consciousness) (Ch.4 of *FPHC*; Ch.10 of *FCD*; & Ch.7 of *FHC*)
  - Ex: New technological forces in material sciences, electronic and communication sciences, energy sciences, biosciences, manufacturing and engineering sciences, and space sciences (Ch.10 of *FCD* & Ch.7 of *FHC*)
  - Ex: Systematic dominance towards nature for space colonization (Chs.9-10 of *FCD*; Chs.2 & 7 of *FHC*)

- *At the Cultural Theoretical Level*
  - Ex: The post-human transcendence of freedom and equality (Ch.10 of *FCD*)

- *At the Cosmological Theoretical Level*
  - Ex: Mastering of dark matter and dark energy, and the exploration of multiverses (Ch.4 of *FPHC*; Ch.10 of *FCD*; & Ch.7 of *FHC*)

- *At Other Levels*
  - Ex: Historical: pre-modernity → modernity → postmodernity → after-postmodernity (human distinction, and the rise of post-humans, including floating consciousness) (Ch.7 of *FHC* & Ch.10 of *FCD*)

(continued on next page)

Table 1.36. Civilizational Holism
(Part II)

Notes: These examples are solely illustrative (not exhaustive), and some of the items can be reclassified somewhere else. Nor are they always mutually exclusive. Since they are generalities, exceptions are expected. And the comparison is relative, not absolute.

Sources: From *Table 5.1* of *FPHC*—with details from *FHC* and *FCD*

Table 1.37. Theories on Civilizational Holism (Part I)

- *At the Biological Theoretical Level*
  - —Ex: Theory of Contrastive Advantages (Peter Baofu)
    (*Sec.2.6* & *Ch.10 of FCD*)

- *At the Psychological Theoretical Level*
  - —Ex: Theory of Floating Consciousness (Peter Baofu)
    (*Ch.10 of FCD*; *Chs.1 & 4 of FPHC*)
  - —Ex: Theory of Cognitive Partiality (Peter Baofu)
    (*Sec.2.4 of FCD*; *Sec.4.5.1.1 of BCPC*)
  - —Ex: Theory of Emotional Non-Neutrality (Peter Baofu)
    (*Sec.2.5 of FCD*; *Sec.4.5.2 of of BCPC*)
  - —Ex: Theory of Behavioral Alteration (Peter Baofu)
    (*Sec.4.5.3 of BCPC*)

- *At the Organizational Theoretical Level*
  - —Ex: Theory of E-Civic Alienation (Peter Baofu)
    (*Ch.7 of FCD*)

- *At the Institutional Theoretical Level*
  - —Ex: Theory of Post-Capitalism (Peter Baofu)
    (*Ch.10 of FCD*; *Chs.2 & 4 of FPHC*; *BCPC*)
  - —Ex: Theory of Post-Democracy (Peter Baofu)
    (*Ch.10 of FCD*; *Chs.3 & 4 of FPHC*; *BDPD*)

- *At the Structural Theoretical Level*
  - —Ex: Theory of the Cyclical Progression of Hegemony (Peter Baofu)
    (*Chs.9-10 of FCD*; *Chs.1, 3 & 4 of FPHC*)

(continued on next page)

Table 1.37. Theories on Civilizational Holism (Part II)

• *At the Systemic Theoretical Level*
  —Ex: Theory of Post-Humanity (Peter Baofu)
    (Ch.7 of *FHC*; Chs.3, & 10 of *FCD*; Chs.1, 3 & 4 of *FPHC*)
  —Ex: Theory of the Cyclical Progression of System Integration and Fragmentation (Peter Baofu)
    (Chs.9-10 of *FCD*)

• *At the Cultural Theoretical Level*
  —Ex: Theory of Methodological Holism (Peter Baofu)
    (Ch.1 of *FCD*; Ch.1 of *FPHC*; Sec.2.1 & Sec.2.5 of *BCPC*)
  —Ex: Theory of the Trinity of Modernity to Its After-Postmodern Counterpart (Peter Baofu)
    (*FHC*; Ch.10 of *FCD*)

• *At the Cosmological Theoretical Level*
  —Ex: Theory of Existential Dialectics unto Multiverses (Peter Baofu)
    (*FHC*; *FCD*; *FPHC*; *BDPD*)
  —Ex: Theory of Hyper-Spatial Consciousness (Peter Baofu)
    (Ch.4 of *FPHC*)
  —Ex: Perspectival Theory of Space-Time (Peter Baofu)
    (*FPHST*)

• *At Other Levels (Historical)*
  —Ex: Theory of the Evolution from Pre-Modernity to After-Postmodernity (Peter Baofu)
    (*FHC*; Ch.9-10 of *FCD*; *FPHC*)
  —Ex: Theory of the Evolution from Barbarity to Post-Civilization (Peter Baofu)
    (*BCIV*)

(continued on next page)

Table 1.37. Theories on Civilizational Holism (Part III)

Notes: All these theories are my constructions, as some of the main contributions of my grant project on civilization and its future. These examples are solely illustrative (not exhaustive), and some of the items can be reclassified somewhere else. Nor are they always mutually exclusive. Since they are generalities, exceptions are expected.

Sources: Based on FHC, FCD, FPHC, BDPD, BCPC, and BCIV

PART TWO
Culture

· CHAPTER TWO ·

Space-Time and Culture

Cultural influences help form a person's view of time [and space].
—Time (1990: 79)

Space-Time and the Influence of Culture

The term `culture´ in the title etymologically originates from Middle English, Middle French, and Latin *cultura*, with its past participle *cultus*, to refer to "the customary beliefs, social forms, and material traits of a racial, religious, or social group" or "the set of shared attitudes, values, goals, and practices that characterizes" the group. (MWD 2005b)

In a more substantive definition, culture, as used in *FCD* (2002: 12), for instance, has in mind "that which concerns the issues of ideology, philosophy, religion, ontological constructs, norms, rituals, folklores, the arts and literature, among other things."

In the present context of space-time, the relevance of culture consists of its influence on the understanding of space-time, to the extent that different values and beliefs about space and time have shaped human life in different ways since time immemorial.

An examination of space-time in relation to culture (as summarized in *Table 2.1*) can therefore be divided into two sections, namely, (2.2) the cultural domains of space and (2.3) the cultural aspects of time, to be respectively addressed below.

It must be stressed, of course, that, as was summarized in *Sec.1.2* on space-time in relativity, space and time do not exist independently but have a relativist interaction with each other.

However, the reason for the separate analysis of space and time in this chapter is solely for the academic convenience of a more detailed examination of the cultural influence on understanding space and time, without committing the same mistake of treating space and time in an absolute meaning (which only brings reminiscence of the now debunked Newtonian absolute perspective of space and time).

Space and Culture

The cultural domains of space can be further be divided into three (illustrative only) categories for consideration, that is, in the absence of better terms, what I want to call (2.2.1) *epistemic space*, from the perspective of epistemology, (2.2.2) *moral space*, from the perspective of morality, and (2.2.3) *aesthetic space*, from the perspective of aesthetics—to be elaborated hereafter, respectively.

Epistemic Space

Seeing an object, for instance, requires a perspective on space whose conception varies from one culture to another, or from a historical era to another.

As an illustration, in Western pre-modern (say, medieval) times, the dominant conception of space often conformed to the three-dimensional Euclidean geometry with its classical causality which treated reality "as a series of scenes on the temporal conveyor belt of sequence." (L. Shlaine 1991: 191)

This "flat, floating arrangements" of reality, with its rather dull Realist depiction, was "rooted itself so deeply in the...[Western] mind [at the time] that no other form of perception could be imagined....The three-dimensional space of the Renaissance is the space of Euclidean geometry. But about 1830 a new sort of geometry was created, one which differed from that of Euclid in employing more than three dimensions." (S. Giedion 1976: 31, 435)

When space is to be understood from the perspective of more than three dimensions by including, say, time, it allows the enriched appreciation of looking at an object from many sides, depending on the multiple angles of observation one is willing to make through time, through self-participation in space.

Sigfried Giedion (1976: 435-6) thus rightly observed: "The essence of space as it is conceived today is its many-sidedness, the infinite potentiality for relations within it. Exhaustive description of an area from one point of reference is, accordingly, impossible; its character changes with the point from which it is viewed. In order to grasp the true nature of space the observer must project himself through it."

Two illustrations of the multifaceted perspective of space can be made in relation to (2.2.1.1) inner and outer space (in transparency), (2.2.1.2) vertical and horizontal space (in relationality), and (2.2.1.3) farness and nearness (in directionality)—to be described below, respectively.

Inner and Outer Space (in Transparency). One excellent way to portray this new perspective of space in cultural understanding is the incorporation of space from both inside and outside, that is, the union of inner space and outer space in modern architectural expressions.

The famous Eiffel Tower in Paris, France speaks volumes of this new union of the two sides of space, in that its long stairways in the upper levels of the tower reflects "the continuous interpenetration of outer and inner space," to the extent that one can see through the outside from the inside, and vice versa. (S. Giedion 1976: 435-6)

At the world-renowned Massachusetts Institute of Technology (M.I.T.) in Cambridge, MA, one of its dormitories, known as Baker Hall and designed by Alvar Aalto, has interior "entrance and projecting staircases when viewed from the outside....As soon as one steps into the entrance one sees through the whole transparent building." (S. Giedion 1976: 637-9)

A major feature of modern architectural expressions is therefore this "extensive transparency that permits interior and exterior to be seen simultaneously," to enlarge the very conception of simultaneous space for cultural/artistic appreciation. (S. Giedion 1976: 493)

Vertical and Horizontal Space (in Relationality). Besides inner and outer space (in transparency), another distinctive expression takes the form of vertical and horizontal space (in relationality). That is, instead of looking from inside and outside of a space, one instead gazes from above and below in relational space.

One good instance is Cubism by Pablo Picasso.

Sigfried Giedion (1976: 435-6), for instance, summarized this Cubist achievement in proposing a new way of appreciating space: "It was in cubism that this was most fully achieved. The cubists did not seek to reproduce the appearance of objects from one vantage point; they went round them, tried to lay hold of their internal constitution. They sought to extend the scale of feeling, just as contemporary science extends its descriptions to cover new levels of material phenomena. Cubism breaks with Renaissance perspective. It views objects relatively: that is, from several points of view, no one of which has exclusive authority. And in so dissecting objects it sees them simultaneously from all sides—from above and below, from inside and outside. It goes around and into its objects. Thus, to the three dimensions of the Renaissance which have held good as constituent facts throughout so many centuries, there is added a fourth one—time."

And the Bauhaus in architecture shares a comparable relational (and also, for that matter, transparent) looking at space: "Two major endeavors of modern architecture are fulfilled here, not as unconscious outgrowths of advances in engineering but as the conscious realization of an artist's intent; there is the hovering, vertical grouping of planes which satisfies our felling for a relational space, and there is the extensive transparency that permits interior and exterior to be seen simultaneously." (S. Giedion 1976: 493)

Farness and Nearness (in Directionality). The German philosopher Martin Heidegger (1962) once introduced different views of space in his major work titled *Being and Time*—along a comparable (though different) angle of observation, this time, in relation to farness and nearness.

For instance, instead of repeating a conventional view of space as an independent "arena" or "container" for objects (that is, "world-space"), he showed how space could be perceived differently in relation to the activities that we do while living in our own world (e.g., "de-severance," "directionality," and "regionality").

In "de-severance" (*Ent-fernung*), as Yoko Arisaka (1996) explained, "[w]hen I walk from my desk area into the kitchen, I am not simply changing locations from point A to B in an arena-like space, but I am `taking in space´ as I move, continuously making the `farness´ of the kitchen `vanish,´ as the shifting spatial perspectives are opened up as I go along."

In "directionality" (*Ausrichtung*), as Arisaka (1996) continued, "[e]very de-severing is aimed toward something or in a certain direction which is determined by our concern....I must always face and move in a certain direction....If I want to get a glass of ice tea, instead of going out into the yard, I face toward the kitchen and move in that direction...."

And in "regionality" (or regions, *Zuhandenheit*), the process goes one step further, in that "a certain direction...is dictated by a specific region." (Y. Arisaka 1996) So, in the above example of getting a glass of ice tea, I end up moving to the kitchen, "following the region of the hallway and the kitchen. Regions determine where things belong, and our actions are coordinated in directional ways accordingly."

Space can thus be understood in relation to nearness and farness, in the context of the activities that we do when living in a specific world.

Moral Space

Besides epistemic space—space also has a moral embodiment in a culture.

A fascinating case study of this moral expression of space in a culture is its architectural space.

Sigfried Giedion (1976: 705) has this understanding well explained in this passage: "Architecture has long ceased to be the concern of passive and businesslike specialists who built precisely what their clients demanded. It has gained the courage to deal actively with life, to help mold it. It starts with intimately vital questions, inquiring into the needs of the child, the woman, and the man. It asks, `What kind of life are you leading? Are we responsible for the conditions you have to put up with? How must we plan—not just in the case of houses, but clear through to regional areas—so that you may have a life worthy of the name?´....Following an impulse which was half ethical, half artistic, they have sought to provide our life with its corresponding shell or framework. And where contemporary architecture has been allowed to provide a new setting for contemporary life, this new setting has acted in its turn upon the life from which it springs. The new atmosphere has led to change and development in the conceptions of the people who live in it."

This should not be surprising, if one only remembers that, especially in pre-modernity, architectural masterpieces had political and moral meanings at times mixed with the sacral realm. For instance, ancient Egyptian pyramids were of course used as a royal treatment for the burials of the pharaohs but, on top of that, contained a theological-moral theme for the reverent union with the cosmological space as expressed through the grandiose peak (in height) of each pyramid.

By the same logic, in many holy places around the world, the theological-moral expression of a religious piety is most noticeable. In the Islamic world, for instance, one of the five pillars of Islam is the pilgrimage to Makkah (the hajj), where reside the holy shrines (together with their magnificent public spaces) for the faithful to experience once in a lifetime the sacral respect towards, and the love of, prophet Muhammad.

Aesthetic Space

From moral space, however, is not far aesthetic space. After all, many artistic works about the nature of space contain both moral and aesthetic meanings.

In the aesthetic realm, although there can be various styles over time, some artistic works often include those on the nature of space which make good use, just to cite a classic example, of the conservative presupposition of beauty on the basis of geometric harmony and precision.

One good illustration is aesthetic neo-classicism. As Lawrence Shlain (1991:85) explained, "artists organized space mathematically, like physicists...and `neo-classicism,´ the term used to describe the works of Jean Auguste Ingres, Jacques Louis David...and others...affirmed the rectitude of rectilinear space and clear, precise logic."

A good illustrative reflection of this aesthetic tendency in treating space is *the golden ratio 2:5* so often used in many artistic works. For instance, it is well known among architects that many gothic buildings follow this ratio in their aesthetic design of space on the outside.

Sigfried Giedion (1976: 687) therefore observed that "[a]rchitecture has always had close contact with the proportions of geometry, regardless of its different forms: the pyramids, the Parthenon, the Pantheon. This has held true for highly geometric forms and for highly organic ones (as in the late baroque) and it is still valid for contemporary architecture....The real secret of Utzon's Sydney Opera house is its obedience to the eternal architectural law: the close relations of architecture and geometry."

Time and Culture

Culture can also have its impacts on the understanding of time—just as it has shaped how we understand space over the ages (as already addressed in the previous section).

It is often noticed by scholars that cultural influences (e.g., mythologies, religious beliefs, philosophical worldviews, and scientific ontologies) shape a person's view of time. (Time 1990: 79)

Three main cultural aspects of time are of interest here, that is, (2.3.1) monochronic and polychronic time, (2.3.2) linear and cyclic time, and (2.3.3) simultaneous and successive time, to be analyzed in what follows, respectively.

Monochronic and Polychronic Time

The distinction between monochronicity and polychronicity lies on a continuum, that is, in a relative sense.

Edward Hall, for one, defined the term `polychronicity´ with such features as "involvement in several things at once," "completion of transactions rather than adherence to present schedule," "appointments not taken as seriously," "time...seldom experienced as wasted," and "time considered to be a point rather than a ribbon or road." (C. Saunders 2005) Monochroncity, on the other hand, is just the opposite, obviously.

In cultures which value more monochronicity, time is therefore treated one-perspectivally, to the extent that the focus is on "fixed hours," "punctuality," "one thing at a time," "rigid timetables and schedules," and the like. (C. Saunders 2005; R. Brislin 2003: 368-371) Excellent examples of cultures with a value saliency on monochronic time include such countries as Germany, England, Nordic Europe, the United States, and Japan. (B. Poole 2000: 380; C. Saunders 2005)

By contrast, in cultures which appreciate more polychronicity, time is touched upon with a flexible sensibility, to the extent that the emphasis falls on "working any time," "unpunctuality," "multiple tasks at one time," "unpredictable timetable," "unexpected change of plans," and so on. (C. Saunders 2005; R. Brislin 2003: 368-371) And good instances of cultures like this are in Southern Europe, the Near East, the Arab World, Northern Africa, and Southeast Asia.

Compare, say, the polychronic time in Latin America with the monochronic time in Western Europe: "Unlike their polychronic counterparts, Westerners strongly believe in what psychologists call closure, the bringing of a task or activity to conclusion before beginning something else. Cultures with a vertical [polychronic] view of time, on the other hand, tend to be quite comfortable with putting aside an unfinished activity indefinitely." (Time 1990: 100)

Consequently, in this light, "Latin American attitudes toward time... reveal the inner workings of the culture. Exasperated North American businesspeople have for years complained about the *mañana, mañana* attitude about Latin Americans. People often are late for appointments; sometimes little *appears* to get done. For the North American who believes that time is money, such behavior appears senseless. However, Glen Dealy, in his perceptive book *The Public Man*, argues that such behavior is perfectly rational. A Latin American man who spends hours over lunch or over coffee in a café is not wasting time. For here, with his friends and relatives, he is with the source of his power. Indeed, networks of friends and families are the glue of Latin American society. Without spending time in this fashion he would, in fact, soon have fewer friends. Additionally, he knows that to leave a café precipitously for an `appointment´ would signify to all that he must not keep someone else waiting—which further indicates his lack of importance." (ISE 2005: 100,106)

In another example, Edward Hall (1969: 32-33) once wrote about the management of time in the Hopi tribe and compared it with its North American counterpart: "We Americans are driven to achieve what psychologists call `closure.´ Uncompleted tasks will not let go, they are somehow immoral, wasteful, and threatening to the integrity of our social fabric....I became increasingly puzzled as I began to realize that the Hopi lacked this crucial concern for closure and that they had no timetables in their heads for ordinary built objects....In those years, one of the most noticeable and arresting characteristics of Hopi villages was the proliferation of unfinished houses which dotted the landscape....Questions by whites as to when the house would be finished were treated as non sequiturs—which they were to the Hopi."

Linear and Cyclic Time

Just as culture can be polychronic or monochronic in relation to time, it can also be linear or cyclic in a different sense.

Both linear and cyclic views of time can be multiple in version. For instance, Western and Bantu cultures share a linear view of time, in that for them, firstly, "time had a beginning, before which time did not exist" and, secondly, "time forward moving." (B.Poole

2001: 376-377) However, their linear views of time differ in a different way, in that the African version (as is the case in the Bantu-speaking culture in central, eastern, and southern Africa), unlike the Western one, treats time as "a revolving sphere that rolls forward along a spiral path in the endless future."

By the same logic, but in the reverse direction, both Hopi and Hindu cultures share a cyclic view of time, in that time does not have a linear beginning as portrayed in Western and Bantu cultures. However, they differ, in that, for the Hindu, time is "a revolving circle of birth, death, and rebirth," whereas, for the Bantu, time is regarded as something of "a unified holistic pattern"; in other words, "the past and the future are blended with, and indistinguishable from, the present. These cultures see time as a fabric with an interwoven pattern of past, present, and future." (B. Poole 2001: 377; F. Melges 1982)

This clarified—the relationship between the two pairs (monochronic vs. polychronic, and linear vs. cyclic) is not completely clear-cut. Yet, there is some correlation, albeit to an extent only.

For instance, cultures with a monochronic time tend, though not in all cases, to be "present-dominated," to the extent that it is oriented towards "the short-term perspective," "time efficiency," and the like—although they can also value future orientation. (C. Saunders 2005; R. Brislin 2003: 374-376; B. Poole 2000: 382-384)

"Present-dominated" cultures highly value efficiency, to the point that it equates time with money, as Benjamin Franklin once famously put it, as an extreme expression of the commoditization of time. (G. Stix 2002:37) Ian Walker of the University of Warwick even went so far as to calculate the money lost for some daily routines in England; for instance, 3 minutes of brushing one's teeth is equivalent to 45 cents spent (in 2002), after taxes and social security paid to the British government; or half an hour washing a car by hand is like $4.50 wasted. (G. Stix 2002: 37)

Even here, there are shades of gray among monochronic cultures; for instance, Kluckhohn and Strodtbeck "compared English and U.S. societies and suggested that Americans [by contrast]...place an emphasis upon the future that is anticipated to be bigger and better," although the Americans are not as long-term as the Chinese towards the future. (S. Beldona 1998; 376) This means that cultural perspectives on time are so relative and differ even in shades of gray.

Similarly, but in the opposite direction, cultures with a polychronic time share an ambivalent duality in perspective. In other words, in one way, they can be more "past-dominated," with its "long-term perspective" tainted with "risk averseness," "emphasis on stability," and so on. (C. Saunders 2005; R. Brislin 2003: 374-376; S. Beldona 1998)

For instance, Edward Hall (1969: 34) had this comparative observation to make about the "past-dominated" culture of the Hopi tribe: "The Spanish priests enslaved the Hopi. Our government didn't do much better....Everything possible was done to destroy the fabric of Hopi life....[W]hite people...were either completely ignorant of the past or else assumed that because `that was ancient history´ the Hopi could not feel intensely about things which happened before those now living were born. Well, the Hopi hadn't forgotten, and for them the past dominated the present."

However, in a different way, cultures with a polychronic time can also be, at times, "future-dominated." As an illustration, Richard Brislin (2003: 376) and Euguen Kim had this interesting observation to say about the Chinese: "Ironically, Chinese are known for their long-term perspective despite their respect for past. They are known for planting slow growing trees for their grandchildren. The traditional practice of lifetime employment in East Asia reflects a long-term orientation in those countries."

The notion of future orientation here is comparable, though to a certain extent only, to the idea of "originary temporality" in a different context by Martin Heidegger, in that "[p]rojection is oriented toward the future, and this futural orientation regulates our concern by constantly realizing various possibilities." (Y. Arisaka 1996)

Now, Heidegger had a different context in mind about originary temporality, not so much in relation to everyday life routines but in the larger context of another view of time, namely, "authentic temporality," which "has to do with one's grasp of his or her own life as a whole from one's own unique perspective. Life gains meaning as one's own life-project, bounded by the sense of one's realization that he or she is not immortal." (Y. Arisaka 1996)

With this clarification in mind—the point here is not to favor one (like the linear view of time) against the other (like the cyclic view of time), nor does this imply that there are only two views of the present and the past (or the future)—but the motif here is to merely show the conflicting views of time in relation to progress.

In the present context, the triumph of "linear-centric" civilization in modernity and its crisis in postmodernity were already analyzed in *BCIV* (and, for that matter, in the two volumes of *FHC*), but this theme will be revisited in later chapters.

Simultaneous and Successive Time

A third comparison on the cultural aspects of time consists of the distinction between simultaneity and succession, relatively speaking of course.

James Gleick, in his book titled *Faster: The Acceleration of Just About Everything*, precisely described how the Federal Express shipping business, as an exemplary indication of the quickening of time in postmodern culture, literally helped creating the new era of simultaneity, when a package could be delivered "absolutely positively overnight." (G. Stix 2002: 37)

Of course, the advent of the Internet has further accelerated the consolidation of this new era of simultaneity, to the point that every piece of news can be known everywhere on earth almost simultaneously. (G. Stix 2002: 37)

And the Swatch Group, the world's largest watchmaker, even proposed a single "universal Internet time" that "divides the day into 1,000 `Swatch beats´" in order to abandon the conventional local time zones in the age of simultaneity. (W. Orlikowski 2002:690)

To keep pace with this ever more simultaneous phenomenon, time clocks are increasingly built ever more precisely, and a good instance is a team from France and the Netherlands who recently built a time clock with a laser strobe light which emitted pulses lasting about 250 *atto*seconds (that is, 250 billionths of a billion of a second). (G. Stix 2002: 39)

Space-Time and the Delimination of Culture

To say that the understanding of space and time in human life is subject to cultural influence should not be misconstrued as falsely claiming that its impacts are absolute.

Nothing is farther from the truth, since the cultural influence has its own limits, as other factors are highly vital in the evolution of understanding space and time.

Good candidates are none other than society, nature, and the mind. So, the next chapter, that is, Chapter Three, is to precisely focus on another major variable in question, namely, space-time and the social factor—to which we now turn.

Table 2.1. Space-Time and Culture

- **Space and Culture**
    - –Ex: epistemic space (e.g., inner and outer space, vertical and horizontal space, far and near space)
    - –Ex: moral space
    - –Ex: aesthetic space

- **Time and Culture**
    - –Ex: monochronic and polychronic time
    - –Ex: linear and cyclic time
    - –Ex: simultaneous and successive time

Notes: These examples are solely illustrative (not exhaustive), and some of the items can be reclassified somewhere else. Nor are they always mutually exclusive. Since they are generalities, exceptions are expected. And the comparison is relative, not absolute.
Source: A summary of Chapter Two.

• PART THREE •
Society

· CHAPTER THREE ·

Space-Time and Society

> The "spatial" view is incomplete, and [there is]...an alternative, more dynamical view [viz., the one on time in society].
> —Jay Lemke (2000: 274)

Space-Time and the Power of Society

The word `society´ in the chapter title stems from its etymological origins in Middle French *societé* and Latin *societat-*, *societas*, from *socius* ("companion"), to refer to "an enduring and cooperating social group whose members have developed organized patterns of relationships through interaction with one another" or "a community, nation, or broad grouping of people having common traditions, institutions, and collective activities and interests." (MWD 2005c)

In a more substantive definition, society, as previously proposed in *FCD*, *FPHC*, *BCPC*, and *BCIV*, falls in the jurisdiction of three main sub-fields in sociology, namely, (a) micro-sociology, (b) meso-sociology, and (c) macro-sociology.

Now, in (a), the topics concern social psychology and will be postponed for analysis until Chapter Four on space-time and the mind.

In (b), the issues focus on social organizations and networks, for instance, and will be addressed in *Sec.3.2* in relation to space-time.

And in (c), the themes belong to the domains of (c1) social institutions, (c2) social structure, and (c3) social systems. They will be analyzed, respectively, in *Sec.3.3*, *Sec.3.4*, and *Sec.3.5* in the context of space-time.

Just as it was earlier clarified in *Sec.2.1* and also in *Sec.1.2* on space-time in relativity, space and time are not absolute and have mutual interactions with each other.

Consequently, the rationale for the separate analysis of space and time in relation to social organizations, social institutions, social structure, and social systems in this chapter (as summarized in *Table 3.1*) has more to do with the academic convenience for a more in-depth analysis of space and time, with no presumption, however, of the old Newtonian fallacy of absolute space and time. This clarification is important, to avoid any hazy misunderstanding.

Space-Time and Social Organizations

The analysis of space-time in the context of social organizations can be further divided, solely for illustrative purpose, into two sub-sections, namely, (3.2.1) space in social organizations and (3.2.2) time in social organizations, to be accounted for below.

Space in Social Organizations

The relationships between space and social organizations can best be summarized as the *spatial arrangements* of social organizations in a society.

Two major illustrations suffice for the present purpose at hand, namely, (3.2.1.1) relational space in organizational management and (3.2.1.2) power space in organizational hierarchy, in what follows.

Relational Space (in Organizational Management). In organization theory, there is a distinction between task-oriented leadership style and people-oriented management style.

Organizations which are paternalistic in orientation are more likely to have what I want to call a close *relational space* among its members than a more impersonal, competitive management style. (P. Baofu 2002: Ch.2)

The relational space within a task-oriented organization tends to be more distant and more likely exists in an individualistic, competitive society (e.g., the U.S. and the U.K.). (C. Saunders 2005) By contrast, the relational space within a people-oriented organization is relatively closer and more likely occurs in a collectivist society (e.g., Japan and China), *ceteris paribus.*

There are some noticeable features in the two styles, though as a conventional stereotype, in that the task-oriented leadership style, *on average,* is congenial to a masculine sensibility for power, wealth, prestige, and the like; whereas the people-oriented management style, *on average again*, reflects the opposite feminine sensibility for personal relationships and, to a certain extent spiritual, concerns (as already described in many of my previous works).

This is so, even though neither the masculine sensibility nor its feminine counterpart refers to men or women exclusively, since the measurement is continuous in a relative sense, not in an absolute one.

This qualification aside—the point here is not to privilege one style over the other but to reveal the multiplicity of relational space in social organizations.

Power Space (in Organizational Hierarchy). By the same tokens, just as there is a relational space in organizational management, there is also something else, or what I want to call the *power space* in organizational hierarchy.

Geerte Hofstede's dimensions of power distance in hierarchical structure is a good case in point, in that, in high power-distance countries (like India, Venezuela, and the Philippines), less powerful members of organizations are more likely to accept their unequal status than their colleagues in low power-distance counterparts (like Israel and the United States). (C. Saunders 2005)

The relatively more acceptance so understood here in the context of high power-distance countries does not necessarily concern the issue of social inegalitarianism in the organizational setting but reveals a more paternalistic tendency in the collectivist culture for the common good—which a more individualistic, competitive organization might do much without.

Since this topic is related to space-time and social structure in *Sec.3.4,* a more elaboration will be given then.

Time in Social Organizations

On the other hand, time also has its own version of organizational shift, that is, the temporal arrangements of social organizations.

Two major issues suffice for illustration here, namely, (3.2.2.1) responsibility time and (3.2.2.2) coordination time, to be addressed below.

Responsibility Time. All social organizations have individuals located within a network, where some have to spend relatively more time in delegating tasks, while others tend to allocate relatively more time to follow the instructions, in what I want to call *responsibility time* in an organization.

Reva Berman Brown (1998) and Richard Herring in "The Circles of Time: An Exploratory Study in Measuring Temporal Perceptions Within Organizations," for instance, talked about the "responsibility code" in a research finding, in which "directors [in an organizational survey]...were more conscious of the difference between work time and private time....The managers were the next highest..., but only 5 % of staff asked which type of planning was wanted....The directors also planned ahead further than the other responsibility levels,...and in their perception of organizational long- and short-term planning. The directors also had the largest measure of useful past when looked at as the number of months they looked back...."

In other words, temporal perception varies from one individual to another, in accordance to the responsibility code (and role) in an organization.

Coordination Time. Time in social organizations also takes a different form in relation to the necessity of ordering and coordinating activities vital to the survival of an organization, and it can be called, in the absence of a better term, *coordination time.*

The main difference between responsibility time and coordination time is that the former has more to do with the allocation of time in accordance to the delegation of responsibilities from the top to the bottom, whereas the latter focuses more on the scheduling of activities in accordance to the functional needs of organizations, regardless of hierarchical ranking, although the two are also related (as will be clear shortly).

For instance, Craig Van Slyke (2004: 22), Douglas Vogel, and Carol Saunders in "My Time or Yours? Managing Time Visions in Global Virtual Teams" thus argued: "The high degree of functional specialization that first emerged during the Industrial Revolution requires the temporal coordination of the many segmented activities within the organization. Temporal coordination requires planning and predicable schedules. Thus, formal organizations need to schedule activities in time, synchronize functionally specialized, time-segmented activities, and allocate the total amount of time among the total set of activities that need to be performed so as to maximize the organization's goals/priorities."

Each of these activities is a challenge on its own right; for instance, scheduling time faces deadlines, synchronizing time requires team rhythms, and allocating time needs performance measures. (C.Van Slyke 2004: 23-4)

All these activities, as shown in the work by Jacqueline Schriber (1987), can ultimately shape the temporal dimensions of an organization, be they about "punctuality" (e.g., "People get upset when you are late for work"), "schedules and deadlines" (e.g., "People here feel that deadlines don't really matter"), "quality vs. speed" (e.g., "It is better to make a bad decision quickly than a good decision slowly"), "synchronization and coordination of work with others through time" (e.g., "Teamwork is not very important here"). (D. Vinton 1992:13) To achieve all of them is no small organizational feat, of course.

At other times, however, there is the danger of over-coordination, which can backfire and become counter-productive. As an illustration, Laurel Goulet (1999: 114) thus observed: "The increased number of hours on the job is often linked to a shift from an industrial era to a knowledge era. Work has largely shifted from a series of tasks to be performed to more open-ended, creative, and demanding tasks that defy standardization. Because they may not be clearly defined tasks to be completed, workers may never be completely finished with their work."

How do managers make sure that the work may never be finished? An answer lies in the research by Leslie Perlow, which "suggests that managers attempt to influence employees to work many hours through three general methods: imposing demands, monitoring, and modeling the behavior." (L. Goulet 1999: 114)

In this sense, there is a correlation between responsibility time and coordination time—although correlation does not mean causation, as this should be read with care.

Space-Time and Social Institutions

Besides social organizations—space-time can also be understood from the perspective of social institutions, be they economic, political, military or else.

Two (illustrative only) categories are at hand for analysis below, namely, (3.3.1) space in social institutions and (3.3.2) time in social institutions, in the following order.

Space in Social Institutions

Two illustrative examples can be offered here, that is, (3.3.1.1) production space and (3.3.1.2) autonomy space, to be addressed below, respectively.

Production Space. In pre-modernity, it was not uncommon, say, for an economic institution to mix labor in production with livelihood in residence into a single space.

But as neighborhoods got larger and villages grew into towns, things became more difficult for the mixture in the economic institution. A separation of the two for what I want to call a more distinct *production space*, in the absence of better terms, was needed.

Sigfried Giedion (1976: 769), for one, well described this transition in the history of the spatial arrangements of the economic institution, when he wrote that "in earlier times the association of production with dwelling quarters was quite natural, but this connection could not be carried over into large towns. Such apartment houses artificially bring together functions which, in an industrial society, should be kept strictly separate. It is absurd in an age of industrial production to permit residence, labor, and traffic to intermingle. It is not merely the endless streets that are inhuman but also the units that go up beside them."

As labor and residence were separated in spatial arrangements, the production space also became more systematically divided to fit in the mass standardization so characterized in modern mass production lines.

Production space in the economic institution was more and more divided into units, especially (though not exclusively) for the service of "capitalist modernity" (as already detailedly described in *FHC*, for instance).

Autonomy Space. As production space develops more fully, there slowly emerges what I want to label as *autonomy space*, that is, to give workers more control of their work in affluent society, as a way to minimize one aspect of the Marxian problematic of work alienation.

P.A. Hancock (1997: 26) in "On the Future of Work" thus wrote: "How can we achieve the goal of designing enjoyment into work? One of the first steps concerns the question of autonomy, or control over working conditions. Control is a vital factor in how work is viewed and how the individual responds to it. Traditionally, industrial workers have had little control over their own activities. In the past, extrinsic control emerged as a function of the manufacturing process itself....This is only one example of how human beings are made slaves....(Servan-Schrieber 1988) However, the injection of some degree of freedom into a person's control over his or her activities accrues important benefits, especially in reducing work-related stress (Karasek 1979)....One classic contemporary example of increasing worker control of work...is...[f]lexibility...reflected in the spatial location of work and...in the structuring of the work by design."

Just as every force encounters a counter-force, there is also a countervailing tendency for more coordination time by management to control over work for ever more unfinished things to do, even in the contemporary era (as already indicated in *Sec.3.2.2.2*).

Time in Social Institutions

Temporal arrangements in social institutions can be subdivided in three areas for illustration, that is, (3.3.2.1) real time, (3.3.2.2) monetary time, and (3.3.2.3) working time, to be described in what follows.

Real Time. A concise illustration is of course the distinction between polychronicity and monochronicity, which was first introduced in *Sec.2.3* on time and culture but is re-introduced here in the context of social institutions.

For instance, some research findings suggest that the larger firms in the economic institution which tend to be more monochronic also value doing things faster. (C. Saunders 2005)

This is especially all the more important in the age of simultaneity, when "the company that survives will be the one that can develop, produce, and deliver products and services to customers than its competitors....Managers in a marketplace characterized by `survival of the fastest´ will need the ability to recognize and deal with temporal forces that exist in any organization and can make the difference between faster and not fast enough." (D. Vinton 1992: 7)

But whether or not this temporal focus on "clockspeed" (e.g., the reduction of some particular activities for the value of something else), especially in the context of doing things faster in "real time," is something desirable (or not) constitutes a value judgment, which is relative to the value saliency of a given firm. (C. Fine 1998; W. Orlikowski 2002)

As an illustration, monochronic firms can experience some forms of alienation, which for some, are regarded as "less humane"; while the trade-off for polychronic firms is its dependency on their leaders. (C. Saunders 2005)

In this light, M. Bennett and P. Weill (1997) even suggested an alternative notion of "real-enough time" to propose a more diverse set of temporal restructuring in accordance to individual needs, to the extent that, as Wanda Orlikowski (2002: 698) and Joanne Yates argued, "such `real-enough´ temporal structures are important…, allowing us to move beyond the fixation on a singular, objective `real time´ to recognize the opportunities people have to (re)shape the range of temporal structures that shape their lives."

Monetary Time. In many social institutions since capitalist modernity, time has also been treated like a commodity, or in the absence of a better term, what I want to label as *monetary time.*

For instance, "in some societies, time is considered as flowing along a linear path, with one event following another, stretching indefinitely into the future. The pace of life is frequently hectic; each day is scheduled, filled with goals to be accomplished in a pre-determined amount of time. Time is equated with money, and idle moments are wasted resources." (Time 1990: 79) This waste of time, for some, constitutes a waste of "leisure time" which could otherwise be spent instead, if not used for productive work. (R. Brislin 2003:371)

Of course, not all societies encourage this kind of social institutions, since they "tend to reflect upon the past and tradition when considering what is to come, rather than rushing headlong about the amount of time it takes to accomplish a given task." (Time 1990: 79)

Social institutions of the latter type are more likely to be less materialistic, or for that matter, less capitalistic (as already analyzed in *FHC, FCD,* and *BCPC,* for instance). In this sense, monetary time can be related to real time (as discussed above).

Working Time. Social institutions can affect time in another different way. In the case of economic institution, the existence of labor unions to enhance the bargaining power over working time is an excellent example.

For instance, Peter Berg (2004: 334-5), Eileen Applebaum, Tom Bailey, and Arne Kalleberg thus showed this relationship in their recent research: "Institutions governing employment relations play a key role in influencing the relative bargaining power of employers and employees over the control of working time. The strength of unions and their position within the employment relations system influence their ability to negotiate work schedules that benefit employees. Strong unions or works councils can be instrumental in monitoring working time at the establishment level and in ensuring that workers are able to take the paid time off that they have accrued when they need it."

On the other hand, weak unions tend to have the opposite effect, in winning fewer concessions from management on working time.

Space-Time and Social Structure

Space-time also has its equivalent in social structure—not just in social organizations (as illustrated in *Sec.3.2*) and social institutions (as in introduced in *Sec.3.3*).

Two separate analyses can be provided hereafter, namely, (3.4.1) space in social structure and (3.4.2) time in social structure, to be elaborated below, in that order.

Space in Social Structure

Space in social structure can in turn be divided for illustration in two main issues, that is, (3.4.1.1) cooperative and competitive space, and (3.4.1.2) discriminatory space, to be addressed hereafter.

Cooperative and Competitive Space. In social structure, there is space for competition, just as there is space for cooperation, or what I prefer to call, respectively, *competitive space* and *cooperative space.*

In relation to competitive space, Anthony Carnevale (1991) in *America and the New Economy,* using the economic realm as the base for analysis, identified four "races against the clock" which many firms compete in today's economy, namely, "first, to develop innovations in technology, products, or work processes; second, to get the innovation `off the drawing board and into the hands of consumers´; third, to increase the efficiency or quality of the innovation or to develop new applications; finally, to use what was learned in the previous steps to move to another innovation." (D. Vinton 1992: 7)

While competitive space can be tight, there is enough room for cooperative space. A good instance is the argument by David Mitrany, the founder of a school of thought known as *functionalism* in international political economy.

Functionalists argue that "as societies and government responsibilities become more complex, government institutions require a greater number of technical (nonpolitical) experts to solve daily problems, ranging from public health to facilitating communication. Given the fact that many of the issues facing modern societies are not confined within national boundaries, international cooperation among technicians, as opposed to politicians, from different countries becomes indispensable." (F. Pearson 1999: 40)

Here then lies the need for cooperative space in social structure too, besides the competitive one.

Discriminatory Space. But cooperative space and competitive space are not the only spaces there be in social structure.

Another excellent illustration concerns discriminatory practices in social structure, or what I want to label as *discriminatory space.* Discriminatory space can be twisted in different ways, in accordance to the factors of age, gender, race, ethnicity, class, region, or whatnot.

In this sense, one should not confuse discriminatory space with either competitive space or cooperative space, since in both competitive and collective social structures there can exist social discrimination, albeit in different ways.

With this clarification in mind—a good example below is deemed sufficient for the present purpose. For instance, in map making for regional geographic comparison, it is well known that the curvature of the Earth renders it difficult for any flat map to accurately represent the true area of the planet. Traditionally, the old Mercator's projector, which was designed back in 1569, is the standard map used everywhere since the expansion of the modern West, as it "gives continents the correct shapes but distorts their relative areas." (J. Gribbin 1994: 41)

An alternative map is the relatively new Arno Peters' projection, drawn in 1974, which "shows continents' sizes in the correct proportions, but it distorts their shapes. Mercator's map shows Europe looking bigger than it really is, which Peters' corrects." (J. Gribbin 1994: 41)

Yet, because Arno Peters' projection correctly shows Europe and North America much smaller than what they misleadingly appear in the Mercator's map, with Africa to emerge as a true giant in the Arno Peters' projection in its correct size, the new map has never been popular in the West.

So, the old Mercator's map with its exaggerated projection of Europe and North America continues to be used in the West, in spite of its misleading distortion in terms of correct size.

Time in Social Structure

Social structure also has its own distortions on the treatment of time, not just on space (as in *Sec.3.4.1* above).

Consider, say, (3.4.2.1) cooperative and competitive time, and (3.4.2.2) discriminatory time, to be analyzed hereafter, in that order.

Cooperative and Competitive Time. Just as there are cooperative space and competitive space in social structure, the same applies to their counterparts in time, or what I prefer to call, in a similar fashion, *competitive time* and *cooperative time*.

In relation to competitive time, each social structure has its own moments of conflicts and, in a more violent way, wars—as is so often seen in human history. The Cold War and the North-South conflicts are two good modern moments of competitive time in international social structure, just as the U.S. invasion and occupation of Iraq since 2003 serve as a most recent instance. (F. Pearson 1999)

And in relation to cooperative time, each social structure equally encounters its own moments of unity and peace—as is often experienced in human history as well. Different regional security alliances (e.g., NATO, SEATO), regional blocs (e.g., NAFTA, EU, APEC), and international organizations (e.g., UN) constitute illustrative moments of cooperative time in international social structure. (F. Pearson 1999)

Discriminatory Time. Just as there is discriminatory space in social structure, there is likewise what I call *discriminatory time* in social structure, in accordance to the basis of age, race, ethnicity, gender, class, region, or else.

Just consider a few instances below.

Firstly, in relation to age and time, the works by R. Kastenbaum (1964 & 1975) and N. Durkee (1964) show that "older people tend to be more past-oriented than younger people." (R. Brown 1998) Similarly, the study by P. Fraisse suggested "that older people perceive of time as moving more quickly than younger people." (R.Brown 1998)

Secondly, in relation to power status and time, the research by Reva Brown (1998) and Richard Herring revealed that "[d]irectors were least tolerant of lateness,...and managers being most tolerant. Finally, directors...talked longest."

And finally, in relation to gender and time, Brown (1998) and Herring also showed that "women look further back into their past, as a percentage of their age, than men....It seems...that women are more cyclic by nature: their monthly menstrual cycle has no male equivalent and has a large effect on their behavior and emotions. Rather more philosophically, linear time presupposes a beginning and an end and carries with it an implication of instability and change: time's arrow points towards the end of the world."

Men, in this linear time, as suggested by Thomas Cottle (1974) and Stephen Klineberg in a different study, are on average more achievement-oriented in their temporal perspective of the future, as compared with women.

Although these examples are by no means exhaustive, they serve the illustrative purpose of revealing discriminatory time in social structure.

Space-Time and Social Systems

Space-time can also be understood at the level of social systems—besides the levels of social organizations (in Sec.3.2), social institutions (in Sec.3.3), and social structure (in Sec.3.4) as described in the past three sections.

There are two sub-sections (which are used for illustration only hereafter) to be divided for analysis, namely, (3.5.1) space in social systems and (3.5.2) time in social systems, in what follows.

Space in Social Systems

Space in social systems can be illuminated in the context of three main categories, namely, (3.5.1.1) urban space, (3.5.1.2) environmental space, and (3.5.1.3) ecosocial space, to be accounted for hereafter.

Surely, there can be more than three categories in the subject matter, but the three examples are used here solely for illustration.

Urban Space. Besides production space in social institutions—a similar effort occurs in social systems, this time, in relation to the organization of what I call *urban space* in towns and cities for better urban planning.

A good illustration is the transition from a two-dimensional to a three-dimensional planning of urban space. By two-dimensional planning, it refers to "the conception of the city built on a single level. The site of the city may be uneven—Rome, city of the seven hills, Greek and Italian hill towns—but in every case the city extended along the ground." (S. Giedion 1976: 862)

Later, both in the medieval and Renaissance eras, as Sigfried Giedion (1976: 45, 52) concisely summarized, "[t]he star-shaped citta ideale of the Renaissance is really the rationalization of a medieval type in which the castle, cathedral, or main square forming the core of the town is encircled by anything from one to four irregular belts. The tree-like plan of Bagnocavallo in Italy shows the organic manner in which a similar situation was handled in the Middle Ages. The difference is that what the Middle Ages brought about organically in a number of different ways the Renaissance proceeded to freeze into a rigid formal pattern from the outset. The medieval city is characterized by expanding belts of streets: the Renaissance, by streets that radiate directly from the center....[T]he citta ideale merely systematized a preexisting medieval type. Both were based on the exigencies of defense."

There came the Baroque perspective, which "was based on a limitless field of vision. Hence typical towns of the late Baroque period, such as Versailles (second half of the 17th century) and Karlsruhe (about a hundred years later), have nothing to do with the star-shaped plan. The palace of the ruler stands boldly between town and country, dominating—at least in the optical sense—limitless space....Michaelangelo was an admixture of Gothic and Baroque. He connects the worldly universality of the Baorque with the spirituality of the Gothic....The architectural significance of the Capitol in Rome can be rapidly summarized. It is a development of Bramanate's use of terraces at the Belvedere into an element of urbanism. It is a comprehensive composition in depth—piazza, stairway, city—and at the same time a preparation for the great axis emanating from a single building, the Senatorial Palace: something the ancient world had never sought to realize." (S. Giedion 1976: 54, 69-70)

By contrast, in three-dimensional planning, the city extends below and above the ground. For instance, "Jorn Utzon's emphasis on the relations of horizontal levels gives a hint in this direction. Urbanism has become the organization of horizontal levels below and above the ground." (S. Giedion 1976: 862)

Cesar Daly in *Revue Generale d'Architecture* revealed the central feature of this three-dimensional planning, be it about "beauty," "action," or "character" of urban space, that is, the obsession with industrial "function," or, in Daly's own words: "By beauty I mean the promise of function. By action I man the presence of function. By character I mean the record of function." (S. Giedion 1976: 216)

Huge metropolis in modern times, with their magnificent skyscrapers, speak thousand words about this new spatial arrangement in action, this time, above the ground.

Environmental Space. Since the fifties, however, a new generation of architects came to challenge the traditional understanding of urban space.

In the process, they proposed a new conception of space for contemporary architecture, that is, the increasing environmental sensibility in urban space, or what I want to simply call *environmental space.*

The new vision incorporates several noticeable features of a quality of life in close connection to nature, beyond the previous obsession with industrial function. (S. Giedion 1976: 668)

Frank Lloyd Wright is a good spokesman of this new era for environmental space. Giedion thus described him as an unusual architect: "Wright...had by nature the will and the courage to protest, to revolt, and to persevere. He carried on in architecture that tradition of sturdy individualism of which in the middle of the last century Walt Whitman and Henry Thoreau were the literary spokesmen....As prophet,...he preached hatred of the city and return to the soil and to the productive, self-sufficient community—in a land where man's relation to the soil is too often remote and impersonal, where at the same moment, according to the varying demands of the economic trend, forests are being changed into farms and acres of growing grain changed back into forests; and where food to a great extent comes to the table out of tins." (S. Giedion 1976: 424)

This reminds us of the older days, when a comparable (though not identical) interaction between nature and architecture was at times emphasized, albeit with an aristocratic accent.

For instance, "[d]uring the century between Versailles (1668-84) and Lansdowne Crescent in Bath (1794) residences came to be placed in direct contact with nature. This direct contact belonged first to the monarch, next to the nobility, then to anonymous wealthy citizens. The increased stress in the eighteenth century upon the connection of the dwelling with nature may be laid in large part to the trend of the times toward Rousseau's cult of the `natural man´....We have observed the interrelations since the

17th century between groups of buildings and nature, and since the 18th century between squares and greenery." (S. Giedion 1976: 711, 754)

Ecosocial Space. Space has another distinctive feature in the context of social systems, especially in relation to the interactions among units in a system, or what I want to label as *ecosocial space.*

A good instance concerns "spherical" topology. In classical systems theory, there is an important assumption, that is, "units nearer in space are more likely to interact and to interact more strongly (i.e., with greater effect on one another). This assumption imposes `spherical´ topology on the system: relative to any center, items at the same distance scale (i.e., in the same spherical shell) are equally likely to be interaction partners, with the closer ones interacting more often or more strongly and the further ones less." (J. Lemke 2000: 274)

But this assumption is now questioned by those studying ecosocial systems as complex ones.

As Lemke (2000: 274; 2000a) explained, in "many complex systems, however, this assumption fails. Two distant points along the same stream may interact more than two nearer points not linked by the system. Two...distant neurons may interact more than close ones not in the same neural network of pathways, sensitized to the same neurotransmitters or neuromodulators. In a pond or an ocean, two species in the same layer of water at the same depth may be more likely to interact over wide (horizontal) distances than they are to encounter a species nearer in vertical distance, but separated ecologically by depth-dependent differences of light, temperature, salinity, or pressure....In our human ecosocial systems,...people who are linked by same river, the same railroad, the same phone network, the same chat room on the internet may interact far more than they do with spatially nearer neighbors who are off these social transport and communication networks. In a modern city, spatial proximity may have little relevance to probability or intensity of interaction."

In fact, the same can be said about the other principles in topology in relation to ecosocial systems, that is, laminar topology (about horizontal layers) and network topology (about lines of connectivity). (J. Lemke 2000: 274)

Time in Social Systems

Time, on the other hand, also interacts with social systems, albeit in a different way.

Three exemplary illustrations suffice here, namely, (3.5.2.1) continuous time, (3.5.2.2) technological time, and (3.5.2.3) ecosocial time, to be addressed respectively in what follows.

Continuous Time. Time has its dynamic interactions with social systems in relation to the past, for instance, or what I want to call *continuous time* in the context of social systems.

A good instance is the work by the third generation in contemporary architecture (around 1950's).

Sigfried Giedion (1976: 670), for one, well summarized the contribution of this group in the history of modern architecture: "Relationship with the past can be both positive and negative. In the U.S. a series of well-known architects of the middle generation has tried to incorporate isolated details and stylistic fragments into their buildings as decorative features. But this selection does not lead to a relationship to tradition or to the past. It leads only to a decadent architecture that delights the public and the press, since it reminds them of the only half-buried ideals of the nineteenth century....A typical example is the Lincoln Center in New York."

So, how exactly do the third generation handle continuous time in a way more distinct from the middle generation?

Giedion (1976: 668-670) thus answered: "The relation of the third generation to the past is expressed differently. It appears in its attitude toward anonymous structures which are everywhere living bonds with the past. The older generation—with certain exceptions—was indifferent to anonymous architecture. It is quite different with the third generation. Wherever one goes one finds a reawaking of the desire to live in a wider span of time; this generation is revolted by the wanton destruction of old buildings in a period of high prosperity....The attitude of the third generation to the past is not to saw out details from their original context. It is more an inner affinity, a spiritual recognition of what, out of the abundance of architectonic knowledge, is related to the present time and is, in a certain sense, able to strengthen our inner security."

It is a mistake, however, to think that continuous time as conceived by the third generation is comparable to the sense of history so often espoused by historians.

Giedion (1976: 670) was quick to clarify this potential confusion: "The attitude to the past of Utzon's generation differs from that of the historian, at least from that of those historians who lack an inner relation to the contemporary scene. The architect is little interested in when or by whom a certain building was erected. His questions are rather: What did the builder want to achieve and how did he solve his problems? In other words, the architect is concerned with searching through previous architectonic knowledge, so that he can immediately confront contemporary architectural aims with those of a former period....The approach to the past always revolves around the same question: How did man in another time under other circumstances solve certain problems, and what were they?"

Technological Time. Time can take another form in social systems in the form of a quickening pace in response to rapid growth in communications technology, or in what I prefer to call *technological time.*

This is all the more so in the globalization process in our times, as Michael Harvey (2001) and Milorad Novicevic rightly put it: "Many managers are experiencing a `quickening´ of their decision-making processes. The globalization of business, the advances in communications technology and increasing demand for prompt data analysis and interpretation have been the primary drivers on the impact of time for decision making."

The dimensions of decision making to be affected include, for instance, "time frame" (e.g., hours, days, months, or years), "tempo" (e.g., "the speed and intensity of interaction events"), "temporality" (e.g., "the limited durability of things, events and processes"), "synchronization" (e.g., the challenge to coordinate business in different time zones), "sequence" (e.g., "a beginning and end to time in a process"), "pauses/gaps" (e.g., "the number and duration of interruptions in a process), and "simultaneity" (e.g., the happening of an event "at multiple locations at the same time"). (M. Harvey 2001)

The distinction between *technological time* in social systems (here in this section) and *real time* in social institutions (as described in *Sec.3.3.2.1*) is that the former focuses more on the technological revolution as its cause in the context of social systems, whereas the latter stresses instead the economic needs behind its pervasiveness in the context of social (or more precisely, economic) institutions.

Ecosocial Time. A third manifestation of time in the context of social systems has to do with the ecology of social formation as a system of time, or what I want to call *ecosocial time*, in short—just as there is *ecosocial space* (as analyzed in the previous section).

Jay Lemke (2000: 278) in "Across the Scales of Time: Artifacts, Activities, and Meanings in Ecosocial Systems" described how the emergent patterns of interactions in a system occur from a lower level to a higher one, which is based on a longer timescale in the process of evolution: "As interactions on some timescale become linked, or coupled, and thus more interdependent, as they do in complex systems..., there are fewer and fewer possible self-consistent patterns....The new cycles take more time, complete on a longer timescale than the individual constituent reactions. They form a new level of organization....Once in place the new cycles also alter the probability of reactions occurring on the level below them, providing downwards constraints. And of course the new level itself now becomes a potential unit of organization for something [like humans] at a still higher level to be built of."

This understanding is important, since it allows us to study "social development not over relatively short timescales (generally the first few years or first decade or two of life)," as is conventionally done, but more fundamentally, on "human projects...over many decades or a lifetime...." (J. Lemke 2000: 287)

So, Lemke (2000: 287-288) concluded by asking difficult questions for future researchers: "We can no doubt learn a great deal from each of these fields about the kinds of projects and activities that occupy the longer timescale spaces on our chart, from those that extend through significant portions of the life of an individual to those that are undertaken by the members of an institution or of a smaller or larger community. What are the longest timescale projects ever sustained by a human community? How should we define the continuity of such projects? What are the means by which integration across timescales is defined in long-range institutional and community projects?"

These are in fact challenging questions to ask.

Space-Time and the Limits of Society

Needless to say, the power of society on the understanding of space-time is by no means absolute, and the examples at each level so far presented in the chapter constitute only some illustrative cases, since there are for sure many more to be discussed.

The point here, however, is only to illuminate the multiple perspectives of space and time from the social horizon that humans have come to construct.

There are other important factors too, besides the variables of culture (as in Chapter Two) and society (in this chapter). The remaining ones to be analyzed concern the mind and nature.

In this light, the next chapter, that is, Chapter Three, is to elaborate on another important factor in question, namely, space-time and the mind—to which we now turn.

Table 3.1. Space-Time and Society

- **Space and Society**
    - *Social organizations*
        - Ex: relational space
        - Ex: power space
    - *Social institutions*
        - Ex: production space
        - Ex: autonomy space
    - *Social structure*
        - Ex: cooperative/competitive space
        - Ex: discriminatory space
    - *Social systems*
        - Ex: urban space
        - Ex: environmental space
        - Ex: ecosocial space

- **Time and Society**
    - *Social organizations*
        - Ex: responsibility time
        - Ex: coordination time
    - *Social institutions*
        - Ex: real time
        - Ex: monetary time
        - Ex: working time
    - *Social structure*
        - Ex: cooperative/competitive time
        - Ex: discriminatory time
    - *Social systems*
        - Ex: continuous time
        - Ex: technological time
        - Ex: ecosocial time

Notes: These examples are solely illustrative (not exhaustive), and some of the items can be reclassified somewhere else. Nor are they always mutually exclusive. Since they are generalities, exceptions are expected. And the comparison is relative, not absolute.
Source: A summary of Chapter Three.

• PART FOUR •
The Mind

• CHAPTER FOUR •

Space-Time and the Mind

> The most violent revolutions in an individual's beliefs leave most of his old order standing. Time and space,...and one's own biography remain untouched. New truth is always a go-between, a smoother-over of transitions. It marries old opinion to new fact so as ever to show a minimum of jolt, a maximum of continuity.
> — William James (BT 2005)

Space-Time and the Impact of the Mind

The term `mind´ here derives from its etymological roots in Middle English, from Old English *gemynd*, related to Old High German *gimunt* ("memory") and Latin *ment-*, *mens*, or *monEre* ("to remind, warn"), which, in turn, is akin to Greek *menos* ("spirit"), *mnasthai*, or *mimnEskesthai* ("to remember"). (MWD 2005d)

It refers to the "complex of elements in an individual that feels, perceives, thinks, wills, and especially reasons" or "the organized conscious and unconscious adaptive mental activity of an organism." (MWD 2005d)

In accordance to my methodological holism (as exemplified in my previous books), space-time is understood here in the context of the mind, that is, in special relation to the three micro levels of the chemical, biological, and psychological.

Consequently, three major themes are of interest here, namely, (4.2) space-time and chemistry, (4.3) space-time and biology, and (4.4) space-time and psychology, to be respectively discussed hereafter (and summarized in *Table 4.1*), all in the context of the mind.

Once more, the qualification which was made in *Sec.2.1* and *Sec.3.1* is to be repeated here too, in that space and time are not absolute but closely interact with each other.

So, the reason for the separate analysis of space and time in this chapter is because of the academic convenience of a more detailed examination of the impact of the mind on understanding space and time, without, however making the false claim that space and time are absolute in their existence (as had already been debunked by the Einsteinian relativist perspective of space and time).

Space-Time and Chemistry

This qualification aside—the analysis of space-time in the context of the mind can be divided into two sections, namely, (3.2.1) space in chemistry and (3.2.2) time in chemistry, to be analyzed hereafter.

Space in Chemistry

Space, in a peculiar way, has a different expression at the chemical level in the context of the mind, or in a way which reflects what I want to call *chemospace*, in the absence of a better term.

Chemospace. An excellent example on the relationship between space and chemistry in the context of the mind is none other than the research finding, which showed that two distant cells may communicate chemically with each other more frequently via the bloodstream than two close ones. (J.Lemke 2000: 274)

Space proximity, then, does not guarantee more mutual interactions among units at the chemical level, or among different cells in the current example about the mind.

This then challenges a fundamental assumption in classical systems theory on "spherical" topology (as already described in the previous chapter and therefore only summarized here), in that "units nearer in space are more likely to interact and to interact more strongly (i.e., with greater effect on one another). This assumption imposes `spherical´ topology on the system: relative to any center, items at the same distance scale (i.e., in the same spherical shell) are equally likely to be interaction partners, with the closer ones interacting more often or more strongly and the further ones less." (J.Lemke 2000: 274)

This assumption now becomes problematic.

Time in Chemistry

By the same logic, time can also be affected differently at the chemical level in the context of the mind.

In the absence of a better term, let me call it *chemotime* (to be addressed below), just as there is *chemospace* (above).

Chemotime. Three good instances suffice for the present purpose.

Firstly, it has been observed that many animals experience "nightly decline in adrenaline and other `awakening´ hormones," which by evolution are to "ensure...that our ancestors would be quiet at night, when they were at greatest risk of harm from predators." (Time 1990: 79)

Secondly, another illustrative instance concerns the familiar "sleep-wake cycle," which "changes significantly" over time; for instance, "people past fifty tend to sleep less at night and nap more during the day than they did when they were younger. It may be [that] the biochemical processes in the brain that affect a person's sense of time also change with age, making time seem to speed by at an ever-increasing clip." (Time 1990: 96)

And lastly, though not least importantly, some studies revealed that "the use of stimulants and psychedelic drugs…can speed up the internal clock of a person." (B.Poole 2001: 379)

So, time at the chemical level in the context of the mind does not stay constant for those who experience it.

Space-Time and Biology

At the biological level in the context of the mind, space-time has its distinctive expressions—just as space-time has likewise shown its own unique forms at the chemical level (as addressed above).

The analysis here can be divided into two parts, that is, (4.3.1) space in biology and (4.3.2) time in biology, in what follows.

Space in Biology

Space takes different meanings at the biological level in the context of the mind.

Biologists over the years notice something fascinating enough, in that different living things use space differently. This constitutes, as a matter of fact, *biospace*, so to speak, to be addressed below.

Biospace. Biospace differentiates different living things in relation to their different spatial niches.

As John.Gribbin (1994: 22) explained, "[t]he amount and type of space individual members of a species need vary enormously from one species to another. Social insects, like ants, live almost literally on top of one another; some large birds, such as albatross, range over vast areas of ocean. Plants cannot move around other than by growing, but they also use space in different ways. Some tiny plants cling to cracks in mountain rocks. Others, like mangroves, thrive in coastal swamps. If two species are in competition, one of them will have to find an alternative niche, or it is likely to become extinct. Wherever there is food, water, and the right temperature, there is life. Life expands to fill the available space."

The same can be said about humans, in that those in China, for instance, with its 1.4 billion population (in the 2000's), have to adapt their lifestyle to their own spatial niches in a way much more different from the situation confronting those in Australia, with its sparse population, on average of course.

Time in Biology

Time also expresses differently in relation to biology in the context of the mind.

In fact, "for more than 3 decades, scientists have been searching for clues to this puzzle [about a master clock for the body's various rhythms and]...recently have...identified a likely candidate—the suprachiasmatic nucleus, a small knot of tissue at the center of the brain." (Time 1990: 87) So, how exactly does this nucleus affect the perception of time in the context of the mind?

In this sense, time in biology in the context of the mind can be distinguished into two main categories for illustration, that is, (4.3.2.1) biotime and (4.3.2.2) ecobiotime, to be addressed below, in that order.

Biotime. Just as there is biospace (as described above), there is biotime too—to be described in what follows.

A few examples suffice for the present illustration.

Firstly, body functions maintain their inner rhythm. For instance, numerous well-documented research findings suggest that "most people isolated from time cues experience a less extreme lengthening of their day—to about 25 hours. But all concluded that without the means to measure time, humans tend to `free-run,' shaping their days according to their own inner schedules. Body functions [are there] to keep to the beat of an inner rhythm." (Time 1990: 78, 81)

In fact, "[m]any living things have a built-in sense of time. People have a natural rhythm of sleeping and waking, and manage their day around the 24-hour spin of the Earth. This body clock actually runs on a rhythm slightly more than 24 hours long, but people can exercise control over it." (J. Gribbin 1994: 20)

Secondly, time perception varies in accordance to the change in body temperature and blood pressure. In fact, neurophysiologist Hudson Hoagland, "cofounder of the Worcester Foundation for Experimental Biology in Shrewsbury, Massachusetts, was the first to discover the connection between body temperature and time perception." (Time 1990: 95)

And finally, the factor of age also has an affect. For instance, a research by Nathaniel Kleitman showed that age "influences the flexibility of a person's rhythms," so "the older a person is, the more rigid his or her rhythms seem to be." (Time 1990: 81)

Indeed, "[t]he natural patterns of biotime also show up in the life spans of different creatures. In general, smaller creatures live their lives more rapidly, and die sooner....Liver cells divide every year or two, while the cells that line the stomach divide twice per day. Cells from a human baby, put in a lab culture, divide 50 times before they die. Cells from a 40-year-old divide only 30 times. This suggests that animals are programmed to have a fixed amount of time to live." (J. Gribbin 1994: 20)

Ecobiotime. At other times, the body changes its time rhythms to get in sync with nature, and this form of time can be what I prefer to call *ecobiotime*, to distinguish it from *biotime* (above), in its mutual interactions with nature.

Just consider a few examples below.

To start, the metabolism of the body can vary in relation to the seasons of the year in nature. For instance, it has been observed by researchers that "the slowing of the body's metabolism in the fall ensured that those same ancestors would have extra fat on their bodies during the cold-scarce winter." (Time 1990: 79)

In fact, "[m]any species, including birds, fish, and land animals migrate over long distances, in tune with the changing pattern of the seasons, maintaining a regular annual cycle....Some species, including tortoises and some bears, can slow their normal life processes so much that they can live through the winter months without eating, in a state of hibernation." (J. Gribbin 1994: 20)

Sometimes, it can also vary in relation to "the alternating periods of daylight and darkness" and "the waxing and waning of the moon," for instance. (Time 1990: 79)

Space-Time and Psychology

At the level of psychology in the context of the mind, space-time possesses its equivalent—not just at the levels of chemistry (as illustrated in *Sec.4.2*) and biology (as in introduced in *Sec.4.3*).

As before, the analysis is to be divided into two sub-sections, namely, (4.4.1) space in psychology and (4.4.2) time in psychology, to be discussed below—both in the context of the mind.

Space in Psychology

Space at the psychological level in the context of the mind can indeed be treated, in the absence of a better term, as *psychospace*, to be accounted for in what follows.

Psychospace. Just as there is biospace, there is also what I want to label as *psychospace*, in relation to the psychological perception of space.

A good case in point concerns different forms of spatial distance in human communication, depending on how close psychologically the participants in question are towards each other. This study is also known as "proxemics" (the study of the use of space) by Edward Hall. (D.Givens 2003)

For instance, Julius Fast (1970:30) distinguished four different forms of spatial distance in interpersonal communication, depending on "how deep or superficial the relationship among the individuals in question is, namely, (a) intimate distance, (b) personal distance, (c) social distance, and (d) public distance, in that descending order of intimacy." (P.Baofu 2005: 135) So, the more intimate the individuals feel psychologically towards each other, the closer the spatial distance among them is in a communication.

Fast (1970: 130, 132) also identified three ways of spatial standing towards each other in communication. Firstly, in "inclusive" vs. "non-inclusive" spatial standing, it reveals how to "plac[e]...the bodies, arms or legs in certain positions." Secondly, in "vis-à-vis [face-to-face] or parallel [side-by-side] body orientation,...face-to-face arrangements indicate a reaction between the two people involved. Side-by-side arrangements, when they are freely taken, tell us the other people are more apt to be neutral to each other, at least in a particular situation." And finally, in "congruence" vs. "incongruence" spatial standing, "congruence of position in a group indicates that all members are in agreement."

The same logic applies at home. For instance, in a "closed family," spatial arrangement reveals that [e]verything about the family is closed in, tight....Everything is in place in these neat, formal homes. We can usually be sure that the family in such a home is less spontaneous, more tense, less likely to have liberal opinions, to entertain unusual ideas and far more likely to conform to the standards of the community." (J.Fast 1970: 135-6)

Time in Psychology

Just as there is psychospace, there is of course what I want to call *psychotime*.

And it can be divided into two categories for illustration, namely, (4.4.2.1) psychotime at the conscious level and (4.4.2.2) psychotime at the unconscious level, to be discussed below.

Psychotime at the Conscious Level. At the conscious level, psychotime can take different forms, be they (a) emotional, (b) cognitive, and (c) behavioral.

Consider an example of each, in what follows.

(a) Firstly, in relation to emotions—emotional changes can affect human time perception. For instance, it is well known among many psychologists that "an individual's psychological perception of time may register an entirely different tempo, fluctuating according to other factors, becoming distorted by events that occur every day. When a person is elated, or concentrating intensely, for example, time seems to fly; but when the same individual is idle, or performing a dull or repetitive task, time drags. And our sense of time's passing changes as we age. Days may seem almost boundless for a child, but they pass all too quickly for an adult." (Time 1990: 78-79)

On the other hand, when a person is depressed, time passes more slowly, "while manic patients experience time moving more quickly." (B. Poole 2001: 379; R.Wyrick 1977)

In other cases, "feelings of intense fear...can expand time, making a split second seem like an eternity. Survivors of automobile and airplane crashes often relate how they felt frozen in time during the brief seconds of terror that preceded their accidents....Scientists believe the close proximity of the planning and timing cells may be the cause of time distortion. When fear strikes, the brain becomes superalert, firing its cells at a terrific rate." (Time 1990: 89)

(b) Secondly, in relation to cognition—cognitive changes likewise make a difference to time perception. For instance, "some experts believe that an increase or decrease in an individual's sensory alertness mobilizes the brain to prepare a response and to screen out all distractions, causing a corresponding change in temporal perception." (Time 1990: 89)

(c) Finally, in relation to behavior—behavioral changes can alter time perception. In fact, some psychologists discovered that "the most extreme example may be the seemingly endless days of a prisoner's life. As inmates adjust to jail life, they learn to fill their time with activities; counting the days until they are released only stretches time further. Sociologist Thomas Meisenhelder notes that `even the language used by prisoners reflects the feeling of effort associated with the passage of prison time.´ Serving a sentence is described in such phrases as `doing time,´ `marking time,´ `putting time´ or `pulling´ time." (Time 1990: 90)

Psychotime at the Unconscious Level. Psychotime can also happen at the unconscious level.

A few illustrations suffice to make this point understood—in relation to (a) behavior, (b) emotions, and (c) cognition, respectively hereafter.

(a) Firstly, in relation to behavior—behavioral predispositions, or more generally called habits at the unconscious level, lead us to behave in a certain way according to some time schedule. For instance, some researchers found out that "habits such as going to bed at the same time each night, rising for work at a set hour each morning, or eating meals at the same intervals each day can serve as cues." (Time 1990:81)

(b) Secondly, in relation to emotions—deep-rooted emotional temperament (such as being an introvert or an extrovert) does make a difference too. For instance, "scientists can...speculate on why some people adjust more easily than others to a shorter or longer day. Curiously, personality type seems to be one factor. Extroverts tend to have more flexible rhythms than introverts. This may be linked to their body temperature cycles. The cycles of extroverts are generally more variable, making it easier to adapt to a new schedule." (Time 1990: 81)

By the same tokens, but in a different way, individuals who are "spontaneous" (that is, "reluctant to plan ahead") are more likely to treat time in macro units than those who are "analytic" (that is, "plan extensively"); so the former "may plan only at the level of things to do this week and rely more on their memory," while the latter "may plan their days in 15 min. or 30 min. intervals, captured in a notebook or some other type of time management device." (J. Cotte 2004: 334; M. Bond 1988; R. Calabresi 1968)

At other times, moods, which are between the conscious and the unconscious levels (that is, at the subconscious level, as detailedly analyzed in Ch.4 of *BCPC*), can be socially contagious in relation to the time perception of a group, as Lucian Conway (2004: 114) argued that "interaction leads to the formation of consensus about mood," that "mood has an impact on time perception," and that "mood consensus leads to consensus in time perception."

(c) And finally, in relation to cognition—cognitive processes can have both the conscious and unconscious forces interacting with each other in affecting human perception of time. For instance, Dutch psychoanalyst Joost Meerloo argued that "our thinking always rushes back and forth in time. The future goal we want to reach directs our course right now, while at the same time unconscious ancient patterns are acted out. This dual temporal determination of both conscious and unconscious intentions points to a complicated relationship. The slogan `The end justifies the means´ overlooks the fact that the very means which are used will always contaminate the end result. Thus the end will not justify the means; influences run both ways in time." (Time 1990: 125-6)

Our perception of time is thus influenced back and forth, consciously and unconsciously.

Space-Time and the Dogmas of the Mind

The impact of the mind in affecting the way that the perception of space-time can be had should not be misleadingly construed as absolute in and by itself.

The reason is that the combined influences at the chemical, biological, and psychological levels so far discussed are solely illustrative, with the reminder that there exist other factors as well, such as the cultural in Ch.2, the societal in Ch.3, and the natural in Ch.5.

The limits of the mind so understood constitute, in this sense, its dogmatic tendency to overrate its importance in understanding space-time.

With this caveat in mind—let's look more closely at the role of nature in changing human perception of space-time, to which we now turn.

Table 4.1. Space-Time and the Mind

- **Space**
    - *—and chemistry*
        - ·Ex: chemospace
    - *—and biology*
        - ·Ex: biospace
    - *—and psychology*
        - ·Ex: psychospace

- **Time**
    - *—and chemistry*
        - ·Ex: chemotime
    - *—and biology*
        - ·Ex: biotime
        - ·Ex: ecobiotime
    - *—and psychology*
        - ·Ex: psychotime (both conscious and unconscious)

Notes: These examples are solely illustrative (not exhaustive), and some of the items can be reclassified somewhere else. Nor are they always mutually exclusive. Since they are generalities, exceptions are expected. And the comparison is relative, not absolute.
Source: A summary of Chapter Four.

• PART FIVE •

Nature

• CHAPTER FIVE •

Space-Time and Nature

[T]here is a...reality, in Nature,...a world of matter, altogether without spirit, life, or mind. This ultimately real world is the world of particles (little bits of dead stuff), of space and time and of forces (gravitational, electromagnetic, and...the strong and weak nuclear forces).
—Jeremy Hayward (BT 2005a)

Space-Time and the Role of Nature

The word `nature´ in the chapter title comes from the etymological roots in Middle English, from Middle French *nature* and Latin *natura*, or *natus*, which in turn is from the past participle of *nasci* ("to be born")—to mean "a creative and controlling force in the universe" or more broadly, "the external world in its entirety." (MWD 2005e)

In the current context, it refers to two levels of physics, that is, the micro-physical and macro-physical worlds in relation to space-time (although other levels are also related, for instance, in the context of the biological, the chemical, and the psychological, as already discussed in the previous chapter and therefore will not be repeated here).

Consequently, the analysis in what follows can be divided into two sub-sections, namely, (5.1.1) space-time in micro-physics and (5.1.2) space-time in macro-physics (or cosmology), to be analyzed hereafter respectively (and summarized in *Table 5.1*).

As is true in the previous three chapters (i.e., *Sec.2.1*, *Sec.3.1*, and *Sec.4.1*), it must be qualified that space and time are not absolute but mutually interactive with each other.

In this sense, the rationale for the separate analysis of space and time in this chapter is because of the academic convenience of a more detailed examination of the role of nature in understanding space and time; there is no presumption here, of course, that space and time are absolute in their existence (as had already been criticized by Albert Einstein in his relativist perspective of space and time).

Space-Time and Micro-Physics

With the above clarification in mind—the analysis of space-time in the context of micro-physics can be divided into two parts, namely, (5.2.1) space in micro-physics and (5.2.2) time in micro-physics, to be analyzed hereafter.

Space in Micro-Physics

Space, in its unique way, has different expressions at the micro-physical level in the context of nature.

Two illustrations are of interest here, namely, (5.2.1.1) quantum-mechanical space and (5.2.1.2) smallest space, to be analyzed in what follows.

Quantum-Mechanical Space. In micro-physics, an excellent illustration of space in action is none other than the peculiar nature of what I want to call *quantum-mechanical space,* in the absence of a better term.

Two features suffice for the present purpose, that is, (a) communication at a distance and (b) translocation.

(a) In regard to communication at a distance, "quantum physicists...have found that nuclear particles `communicate´ at a distance, without having any recognized `channels of sensation.´" (Stuttman 1992: 126) This used to be controversial in 20th century physics, but, by the early part of the 21st century, physicists are less intolerant about the view.

(b) More interestingly, quantum physicists also discovered that the particles "indulge in `translocation,´" meaning that "they can move from one orbit to another without, apparently, traversing the intervening space." (Stuttman 1992: 126)

Although one can be tempted to claim "that what happens at the level of the electron or neutrino is not necessarily applicable to human beings," it is no longer necessary, at least, "that communication at a distance and translocation are contrary to the laws of nature." (Stuttman 1992: 126)

Yet, this has much inspired some science-fiction writers to speculate about future spacefarers who may come up with a technology to move intelligent life in daily life from one part of space to another without traversing the intervening space.

In fact, some quantum physicists already took a very preliminary step towards this far-stretched technological feat in the distant future. For example, the group led by quantum physicist Pan Jianwei at the University of Science and Technology of China already successfully tested a "30-40 km secure quantum communication" for the current world's longest secure "free-space quantum teleportation," as published in the British journal *Nature* issue of July 1, 2004, which is "a breakthrough in quantum computation and quantum communication network." (PD 2005)

But whether or not, or how far, this first step towards quantum teleportation will eventually lead to the technological revolution of moving intelligent life in daily life from one part of space to another without traversing the intervening space remains of course to be seen in distant future history—in light of the relatively technologically primitive era of ours.

Smallest Space. Another insightful feature is the observation by Max Planck on the shortest unit of space, or what is now known as the Planck's unit of length.

According to Planck's theory, "each quantum of radiated energy (photon) has a certain wavelength, and the amount of energy in the photon is inversely proportional to its wavelength. His formula was able to describe the amount of radiation at each wavelength coming each second from 1 square centimeter of a so-called black body (an ideal scientific construct that absorbs or emits radiation of all wavelengths), which emits radiation only as a function of its temperature." (N. McAleer 1987: 219)

From there, he therefore found that "there were natural or standard units for length, time, and mass that allowed the equations of physics to take particularly simple forms. All these units were inconceivably small....But the time and length units were far, far smaller than anything that could have meaning in terms of the physics of the turn of the century. The Planck unit of length was less than billionths of trillionths of trillionths of an inch." (N. McAleer 1987: 219)

This constitutes then the smallest space possibly measured in nature. But what is interesting here is that "Planck's quantum length or distance (trillions of times less than a trillionth of an inch) has even more significance," as will be described later in *Sec.5.3.1* on space at the macro-physical level.

Time in Micro-Physics

Time also takes some unconventional forms at the micro-physical level.

Consider, for illustration, two instances, namely, (5.2.2.1) shortest time and (5.2.2.2) quantum-mechanical time , to be analyzed in what follows.

Shortest Time. Just as there is smallest space (as in *Sec.5.2.1.2*), there is also shortest time at the micro-physical level.

Again, Planck's constant is to be used here, in that the smallest space as measured by the smallest possible wavelength (which has the highest possible amount of energy) has its shortest possible unit of time for measurement.

Consequently, just as the there is the "Planck unit of length" which is "less than billionths of trillionths of trillionths of an inch," there is the corresponding shortest time for that smallest space, that is, "Planck's unit of time, 10-43 second....This Planck time, beyond which Einstein's General Theory of Relativity cannot predict conditions, is usually considered to be the beginning time of the Big Bang, the birth time of our Universe," as will be further elaborated in *Sec.5.3* on macro-physics. (N. McAleer 1987: 219)

Quantum-Mechanical Time. Just as there is quantum-mechanical space (in *Sec.5.2.1.1*), there is what I call *quantum-mechanical time*, in the absence of a better term.

One fundamental feature about quantum-mechanical time is that things can prop up in different spaces at the quantum-mechanical level so quickly that it is hard to tell which of two events constitutes the cause (or the effect), or whether or not they happen simultaneously.

The very distinction among past, present, and future comes into question. Since this question is related to the topic on different philosophies of time, it is postponed for further analysis until *Sec.5.3.2.2* on time-reversal variance.

Space-Time and Macro-Physics (Cosmology)

By contrast, space-time at the macro-physical (or cosmological) level has its distinctive expressions—just as space-time at the micro-physical level shows its own versions of difference.

Again, the analysis here can be divided into two sub-sections, that is, (5.3.1) space in macro-physics and (5.3.2) time in macro-physics, in what follows.

Space in Macro-Physics

Space in cosmology can be illustrated by two amazing features at the macro-physical level, namely, (5.3.1.1) multi-dimensional space and (5.3.1.2) empty space, to be addressed below, in that order.

Multi-Dimensional Space. It is interesting to note that the smallest space at the micro-physical level is closely related to what I call *multi-dimensional space* at the macro-physical level, in the absence of a better time.

For instance, as described in *Sec.5.2.1.2*, at the mirco-physical level of smallest space, something peculiar happens. Neil McAleer (1987: 220-1) thus correctly observed: "Down to that range of distance, space was smooth and even. But at even shorter lengths, it ceased to have three dimensions. Space had, instead, ten dimensions. The others were rolled up, curled up within Planck's distance."

This is all the more so in space-time prior to the Big Bang. For instance, "today's theoretical physicists such as Stephen Hawking and John Wheeler have even attempted to understand what happened before the Big Bang, before Planck time. One such theory states that before the Big Bang, before the birth of our Universe, time and space were very different from the smooth, continuously flowing media of our experience. Because there was nothing with which to measure distances or intervals of time, we could say that everything was happening all at once, and in the same place....To understand space and time before the Big Bang,...[n]ew theory has it that, at any moment, any point in space could spontaneously become a microscope black hole, with a mass given by Planck's value unit for mass, equal to a small grain of dust. That black hole would have a radius given by Planck's extremely small value for length." (N. McAleer 1987:220-1)

Even then, a black hole of this sort is inherently instable and exists in a violent state of nature: "But the black hole would not live long; it would rapidly evaporate back into the space from which it formed, and it would do this in Planck's time....At that minute scale, space and time were violently active things, continually casting up mini black holes, dissolving them, curling away into seven extra dimensions, reconnecting and forming new black holes. This is called the foamlike structure of space-time." (N. McAleer 1987:220-1)

But this only raises a more difficult question, to be addressed below.

Empty Space. The further question can now be asked here: Where do new mini black holes like this (afore-described) come from? A rather mind-boggling answer is that it can come from the void, out of nothing at that level of space.

For instance, Neil McAleer (1987: 222) thus explained: "There is a law that says particles can arise this way as long as they disappear rapidly....The way this works in physics is that a mini black hole, having the infinitesimal Planck mass, can live no longer than the Planck time before it must return to the void. Lighter particles, arising out of nothing, can live for longer times, and a particle of zero mass can live forever."

This view may seem strange indeed, but in Eastern cosmologies, such a view is no stranger, since "[i]n the predominant Eastern philosophies, however, empty space was the void. In Zen teachings, this plenum contained within it the pregnant possibility of everything. From this invisible cornucopia issued forth all that was substance. The large empty spaces contained within an Asian work of art are a representation of this idea. In contrast to a homogenous Euclidean space that never changes, the Eastern view suggests that space evolves. In the one, space is dead and inert, in the other it has organic characteristics. To the scientist working in the 19th century, the idea that empty space was an invisible generative living tissue was fanciful, childlike, and not to be taken seriously." (L. Shlaine 1991: 160-1)

In this light, then, Leonard Shlaine (1991: 160-1) therefore concluded: "[I]t came as a surprise, therefore, when early 21st century Western scientists discovered that particles of matter can in fact be wrung out of a seemingly empty field by quantum fluctuations. From out of a desertlike vacuum can come a squirming proliferation of inhabitants from the particle zoo. This confirmation of the ancient Eastern idea that empty space is alive and procreative forced a reluctant West to rethink its ideas about space. Eastern conceptions of space turned out to be closer to the truth than the flat angular sterile space of Euclid."

It will be some time, indeed, before the notion of empty space is ever more elaborated.

Time in Macro-Physics

Time at the macro-physical level is not less weird, from the perspective of daily life.

Consider, say, (5.3.2.1) imaginary time and (5.3.2.2) reversible time, in what follows.

Imaginary Time. Imaginary time has something to do with the event horizon of a black hole, in relation to the deeper interior of the singularity point.

As Leonard Shlaine (1991: 360) explained, "[i]nside the event horizon [of a black hole], time, like space, ceases to exist in the sense that we know it. Instead, mathematicians have speculated that another type of time exists there: imaginary time." What then is this so-called imaginary time?

Imaginary time has some peculiar features, unlike conventional linear time. For instance, cosmologists "propose that this mental construct is positioned at right angles to the rectilinear arrow of proper time. If time can indeed have another direction that is at right angles to linear time, then time is implicitly spacelike. Furthermore, a two-dimensional, timelike world in space implies a third perpendicular to the right angle of time....As time gains dimensions on the far side of the event horizon, so, too, space by contrast loses them. Inside the event horizon there are limits on breadth and depth, but none on length. Movement sideways or back and forth is restricted: All movement must go inevitably forward toward the singularity. Outside there are three vectors of space and only one relentless direction of time. Inside the event horizon, time opens like an umbrella to contain other vectors while space inexorably is reduced to one, funneling into the singularity." (L. Shlain 1991: 360)

Needless to say, imaginary time so understood is alien to common sense in ordinary daily life.

Reversible Time. Another more fascinating topic about time is the question concerning whether time can be reversed or not.

This topic about time reversibility (or time-reversal variance), or in the absence of a better term, *reversible time*, is one of the most important symmetries of physical laws under debate in the community of physicists and philosophers.

Or put in a more technical way, for time to be irreversible, "in the quantum-theoretic context, it is transition probabilities between reversed states which must equal the probabilities of the unreversed states taken in opposite temporal order for the laws to be time-reversal invariant." (L. Sklar 1974: 368)

The evidence so far is not very clear. Lawrence Sklar (1974: 372), for one, summarized a general view in out time: "Time-reversal invariance has also been found by some experimenters to be violated in some elementary particle interactions, but here the experimental evidence is much less clear cut than it is in the `violation of the conservation of parity,`" for instance. (L. Sklar 1974: 372)

This then raises the further question of whether the distinction among past, present, and future is real or not. Albert Einstein, for instance, thought that the temporal distinction is an illusion. (C. Pickover 1998:6)

As Clifford Pickover (1998: 11) explained, "[t]he distant event jumps either in the future or past depending on whether you walked toward or away from the distant galaxy. The time order of an event on Earth and on another galaxy can be reversed. However, for this to happen, the two spatially separated events must occur sufficiently close in time so that light (or any signal) does not have time to get from event to the other. As a result, there can be no causal connection between the two events because no information or physical influence can travel faster than light between the events. This means that the time order of two events can be changed at whim simply by ambling about, but you can't reverse cause and effect, producing causal paradoxes. For example, different observers may disagree about whether Mr.Veil's laser beam reached the two ends of the ship at the same time, or if the magazines are hit before the newspapers; but all observers agree that the blasts from the lasers leave the lasers before they arrive at the newspapers and magazines....What the relativity of simultaneously means is that events in the past and future are `real´ as events in the present."

Other thinkers, of course, may disagree on the philosophy of time. Indeed, there are two opposing philosophies of time here: "One theory suggests that the flow of time from the past to the future is genuine, and only the moment `now´ is real at any one time. This seems to mean that we need another `layer´ of time in which to measure the `flow´ of our time. If our movement through time implies the existence of a `suppertime´ to measure the rate time passes, there must also be a `super-supertime´ to measure suppertime, and so on. This idea was proposed by 20th-century British philosopher John William Dunne." (J. Gribbin 1994: 38)

The second possibility, on the other hand, suggests "that everything that has happened and everything that will happen exists somewhere in four-dimensional spacetime. This means that, in theory, every moment coexists, in a spread-out spacetime reality. All that `moves´ is our perception of `now.´ One version of this has been proposed by the British astronomer Sir Fred Hoyle. The problem is that this seems to leave us with no options for making choices, that is, it means we have no free will." (J.Gribbin 1994: 38)

Surely, our concern here is not so much with the question of free will, but with that of reversible time.

Space-Time and the Contingency of Nature

If nature interacts with space-time in the different ways as analyzed in this chapter, it is contingent enough, since other factors also play a major role in understanding how space and time have been hitherto understood.

Other important ones were already discussed in previous chapters, especially in relation to Chapter Two on space-time and culture, Chapter Three on space-time and society, and Chapter Four on space-time and the mind.

Of course, one should not deny the factor of luck (or randomness), but it is already embedded in each of the four categories in question (that is, culture, society, the mind, and nature).

But this academic journey is no idle scholarly exercise, since a most important question concerns how all these understandings of space-time help us predict future history, especially in relation to post-humans and their future in heaven and earth?

This then is the central question, to which we now turn to Chapter Six for more analysis.

Table 5.1. Space-Time and Nature

- **Space**
    - *—and micro-physics*
        - ·Ex: quantum-mechanical space
        - ·Ex: smallest space
    - *—and macro-physics (cosmology)*
        - ·Ex: multi-dimensional space
        - ·Ex: empty space

- **Time**
    - *—and micro-physics*
        - ·Ex: quantum-mechanical time
        - ·Ex: shortest time
    - *—and macro-physics (cosmology)*
        - ·Ex: imaginary time
        - ·Ex: reversible time

Notes: These examples are solely illustrative (not exhaustive), and some of the items can be reclassified somewhere else. Nor are they always mutually exclusive. Since they are generalities, exceptions are expected. And the comparison is relative, not absolute.
Source: A summary of Chapter Five.

PART SIX
Conclusion

· CHAPTER SIX ·

Conclusion: Space-Time and Post-Humans

Man...must go on, conquest beyond conquest. First, this little planet and its winds and ways. And then all laws of mind and matter that restrain him. Then the planets about him. And at last, out across immensity to the stars. And when he has conquered all the deeps of space and all the mysteries of time, still he will...live and suffer and pass, mattering no more than all the other animals do or have done. It is this, or that. All the universe or nothing. Which shall it be, Passworthy? Which shall it be?

—Herbert George Wells (BT 2005b)

The Post-Human Challenge

The word `post-humans´ here is a neologism that I originally coined in *FHC*, which was later further elaborated in subsequent books of mine (such as *FCD*, *FPHC*, *BCPC*, and *BCIV*, just to cite some examples).

For instance, in *FCD* (89), I wrote: "As addressed in Ch.7 of *FHC*, a later epoch of the age of after-postmodernity (that is, at some point further away from after-postmodernity) will begin, as what I called the `post-human´ history (with the term `post-human´ originally used in my doctoral dissertation at M.I.T., which was finished in November 1995, under the title *After Postmodernity*, still available at M.I.T. library, and was later revised and published as *FHC*). The post-human history will be such that humans are nothing in the end, other than what culture, society, and nature (with some luck) have shaped them into, to be eventually superseded by post-humans (e.g., cyborgs, thinking machines, genetically altered superior beings, and others), if humans are not destroyed long before then. The post-human history will therefore mark the end of human history as we know it and, for that matter, the end of human dominance and, practically speaking, the end of humans as well."

In light of the afore-analysis of space-time from the multiple perspectives of culture, society, the mind, and nature—a profound question to ask thus concerns the future of space-time in the post-human age.

In *FHC*, *FCD*, and *FPHC*, for instance, one of my main arguments is that the post-humans, especially in the more distant stage of post-human history, will have to colonize in deep space and expand through this universe and beyond for the search of different hegemonies among themselves.

And one noticeable aspect of this distant future which is crucial to the evolution of this post-human history is the mastering of space-time, to the point of altering space and time for hegemonic expansion in deep space—be it in the form of stretching/shrinking space-time, engineering more dimensions of space-time, manipulating different multiverses, or something else which is too far beyond our current scientific understanding.

The Post-Human Alteration of Space-Time

The post-human alteration of space-time in the distant future of post-human history may sound preposterous in our relatively technologically primitive age, but it is essential for space colonization in the ultimate conquest, by intelligent life, of the universe and beyond.

More specifically, there are some main reasons (to be used hereafter for illustration only) behind this future alteration of space-time, namely, (6.2.1) the need to make new matter-energy, (6.2.2) the need to create new space-time, and (6.2.3) the need to conquer the cosmos unto multiverses—to be discussed in what follows, respectively (and summarized in *Table 6.3*).

It should be stressed, however, that the three reasons are all related, in that they all contribute to the evolution of intelligent life in the cosmos unto multiverses in the most distant future beyond our current knowledge.

The Need to Make New Matter-Energy

One cause for the need of altering space-time is the demand for enormous energy supply for future space colonization across the cosmos by intelligent life in the post-human era at some distant point of after-postmodernity.

Since the second law of thermodynamics imposes some constraints on the availability of free energy, humans are currently living in a declining universe. What then is the solution?

There are multiple suggestions, as I wrote in *FCD* (2000: 465): "Surely, this limit of energy supply might not happen, since post-humans will likely develop to the point of one day `rearranging galaxies and saving them from collapsing,´ as Freeman Dyson suggested, or alternatively, because a new Big Bang might occur in any of those random quantum fluctuations in Hawking radiation (that is, as the galaxies use up `the interstellar gas and dust´— from which `new stars condense´—and then collapse in black holes which `exude energy´ in random bursts of `particles and radiation´ before eventually exploding and vanishing). (D. Overbye 2002:D7) Others suggested that perhaps these advanced civilizations will find a way to build gigantic spheres to trap any energy and light in their solar systems from escaping into outer space. (SCICH 2002b) And Alan Guth (1997:248) even proposed different `pocket universes,´ such that `[w]hile life in our pocket universe will presumably die out, life in the universe as a whole will thrive for eternity. ´"

So, as I continued, "perhaps one day the post-humans would figure out a way to migrate to these other `pocket universes´ and to manipulate the physical laws in those universes for their own survival. Or they might even create a new pocket universe in the laboratory, and all it takes is to create a `desired false vacuum´ `by heating a region of space to enormous temperatures´ and `then rapidly cooling it,´ for it to eventually expand exponentially, but the drawback here is that the probability of doing it successfully is extremely unlikely, so it may be better to depend on `natural´ process of eternal inflation." (P. Baofu 2002: 465-6)

More interestingly, recent theories in cosmology also proposed that new universes may ever be born all the time: "Now imagine what might happen if two...bubble universes touched. Neil Turok from Cambridge, Burt Ovrut from the University of Pennsylvania and Paul Steinhardt from Princeton believe that has happened. The result? A very big bang indeed and a new universe was born—our Universe. The idea has shocked the scientific community; it turns the conventional Big Bang theory on its head. It may well be that the Big Bang wasn't really the beginning of everything after all. Time and space all existed before it. In fact Big Bangs may happen all the time. Of course this extraordinary story about the origin of our Universe has one alarming implication. If a collision started our Universe, could it happen again? Anything is possible in this extra-dimensional cosmos. Perhaps out there in space there is another universe heading directly towards us—it may only be a matter of time before we collide." (BBC 2005) Thus is the creation of new energy in a new Big Bang.

But how is the creation of new energy related to the creation of new matter—and, for that matter, to space-time?

Two issues are to be addressed hereafter, that is, (6.2.1.1) matter-energy and (6.2.1.2) space-time.

Matter-Energy. As proposed in the special theory of relativity (1905), matter and energy are interchangeable, with the famous equation, $E = mc^2$. So, if it is possible to create new energy, it is also possible to convert it into new matter. (L. Shlain 1991: 325)

In other words, just as matter can be converted into energy, and the explosion of a nuclear bomb is a most powerful expression of that—the reverse also applies, in that matter can be made out of energy in the nothingness of the void, as was the formation of elemental particles in the beginning of the Big Bang. (L. Shlain 1991: 325)

Alan Guth, the pioneer of inflationary theory in cosmology at M.I.T., once beautifully put it: "It is often said that there is no such thing as a free lunch. The universe, however, is a free lunch." (P. Davies 1983:216)

In fact, even in our relatively technologically primitive era (by future standards), physicists are already busy in creating new forms of matter in different ways—although in extremely modest ways by comparison, that is, in ways which are not as grand as the one that matter was created in the beginning of the Big Bang.

For example, the conventional classification of matter (into liquids, gases, solids, and plasma) is no longer complete, as some scientists at MIT created a new form of matter (the "superfluid"—"a gas of atoms that shows high-temperature superfluidity") in 2005 and another new form of matter (the "Bose-Einstein condensates"—"a new form of matter in which particles condense and act as one big wave") in 1995. (L. Valigra 2005) Later, in 2006, they found a way to observe the superfluid directly. (MIT 2006)

In 2004, another group of physicists, this time, at the National Institute of Standards and Technology in collaboration with the University of Colorado, succeeded in creating another new matter form called the "fermionic condensate," also based on the earlier creation of Bose-Einstein condensates. (CNN 2004)

More recently, in late 2005, physicists from the Beijing-based Institute of High Energy Physics (IHEP), the Chinese Academy of Sciences (CAS) and the University of Hawaii discovered a new matter (a new sub-atomic particle) known as "X1835," as published in December 31 edition of *Physics Review Letters*, one of the world's most prestigious journals for physics research. (PD 2006)

The new particle has a mass "slightly less than twice the proton mass. Its lifespan is also very short, about 10-23 seconds....Almost all known particles that experience strong nuclear force are composed of either two or three quarks, the smallest known material. Yet particle physicists have long predicted the existence of other types of particles, including those containing more than three quarks, or those made of gluons and quarks." (PD 2006)

And X1835 is a good candidate of this other type of particles, except that it is not yet fully understood, so the "X" in the name reveals this ignorance.

Yet, the importance of the discovery should not be ignored, as Tord Ekelof, "a professor at Uppsala University in Sweden and a leading international physicist," explained that "[t]his could be the first evidence that matter and anti-matter can be bound together to become a new form of matter. It is a very important discovery in physics." (PD 2006)

In fact, a recent experiment by the Collider Detector at Fermilab (CDF) international collaboration at the Fermi National Accelerator Laboratory further showed that it is possible to mix both matter and their anti-matter counterparts: "Known as mixing, this process has been known to quantum physicists for 50 years. Now it has been measured for the first time by an international collaboration involving MIT scientists....The CDF team specifically reported rapid-fire transitions between matter and antimatter of a subatomic particle called the Bs (pronounced 'B sub s') meson. They found that this particle oscillates between matter and anti-matter states at a mind-boggling 3 trillion times per second. The Bs itself is composed of other subatomic particles: a heavy 'bottom quark' bound to a 'strange anti-quark.'" (D.Halber 2006)

Yes, the world of the forms of matter is getting more and more crowded, as there are now liquids, gases, solids, plasma, Bose-Einstein condensates, fermionic condensates, X1835 particles, and even Bs mesons.

But this is only the beginning of many more discoveries to come.

Space-Time. In the process of creating matter-energy, the emergence of new space-time is closely related—with the current context on the need for new matter-energy as an illustration.

•CONCLUSION: SPACE-TIME AND POST-HUMANS• 169

And the idea of new "pocket universe" (as suggested by Alan Guth at MIT in the previous section) is solely used as an illustrative example here (since there will be in the distant future many more sophisticated ways of creating space-time that our current world has never known).

After all, in general theory of relativity (1915), matter-energy is unified with space-time, to the extent that any major change of matter-energy will have an impact on space-time, and the reverse also holds true.

Post-humans will most likely find some most advanced ways then to manipulate physical laws to survive in those new universes (as this is related, however, to the issue in *Section 6.2.3*, right after the next section, on conquering the cosmos unto multiverses).

The Need to Create New Space-Time

All this, in a most important way, is to make it as much easier as possible for post-humans to colonize in deep space, which will then require enormous amounts of matter-energy.

But there is something else highly important to remember, that is, another major reason to alter space-time in the long haul.

In other words, space colonization also requires the ability to travel through vast distance in the cosmos. I thus wrote in *FCD* (2002: 466): "Just imagine how long it would take to travel throughout the cosmos, where several hundred billions of galaxies exist, with several hundred billions of stars in each. Our nearest galaxy to the Milky Way, Andromeda, is already some 2 ½ million light-years away. (A. Guth 1997: 214) And our galaxy alone, the Milky Way, is 100,000 light years in diameter, and each light year is 10 trillion kilometers in distance, even if measured at the incredible light speed of about 300,000 kilometers per second in empty space. (CSM 1979)" (P. Baofu 2002: 466)

With this long distance in mind, some scientists like Joachim Hoizer, John Kelvin, and Nikolai Kozyrev imagined some spaceships to travel fast in space, and a good proposal is that "a fast-spinning circle combined with a ring-shaped magnet in a strong magnetic field can 'push' a space ship to other dimensions where different values of the natural constants, including the speed of light, may exist. The machine will be capable of creating antigravitation by moving a spaceship in regular space." (PR 2006)

Another way to overcome the challenge of huge distance in space is to alter space-time in a different way, as already suggested in *FCD* (2002: 461): "To save time and energy in interstellar space travel, some even proposed to go around the ultimate limit of the speed of light by traveling in `warp drive´ (as in science fiction), that is, to `stretch´ and `shrink´ space from behind and in front of a spacecraft, as the physicist Miguel Alculierre proposed. (DWC 2001...) Of course, this is an enduring science fiction dream yet to be realized, because of our current technological backwardness (e.g., how to come up with enormous energy for a spacecraft to stretch and shrink space, and to protect it in the process from being smashed). Others instead proposed the use of `warm holes´ to cross galaxies in little time. (SCICH 2002...)" (P. Baofu 2002: 461)

To alter space-time is thus necessary for this additional reason (as already summarized in *Table 4.6* and *Table 4.7* of *FPHC* and here in *Table 6.1* and *Table 6.2*).

The Need to Conquer the Cosmos Unto Multiverses

Yet, there is still another reason for the alteration of space-time, namely, the continued exploration of multiverses for the ultimate dominance by intelligence life in the most distant era beyond the furthest stretch of our current imagination (as already discussed in *FHC* and *FCD*—and summarized in *Table 4.6* of *FPHC* and here in *Table 6.1*).

As an illustration only, by 2006, for instance, scientists in the Mars Gravity Biosatellite Program at MIT already start working on the so-called "inner solar system economy," in which "they expect humanity to spread out to the moon and Mars to colonize and exploit the resources of these new frontiers." (M. Baard 2006)

Although this only constitutes a small step towards the extremely long evolutionary process to conquer the cosmos unto multiverses, the point here is that space colonization is far from being only a science fiction but is increasingly gaining currency as a serious scientific enterprise, with enormous implications for the future post-human alteration of space-time.

Besides, as discussed in *FPHC*, there is another rationale for this post-human alteration of space-time, in that human consciousness will not last either and will take different post-human forms.

And a good illustration of this is none other than what I originally called the *hyper-spatial consciousness* which allows existence in different dimensions of space-time not currently accessible to humans who are accustomed to the four-dimensional world in this relatively primitive point of history.

When this alternation of space-time happens, it will constitute another milestone in the evolution of intelligent life in the universe and beyond unto multiverses.

Table 1.4 summarized some of the major causes of the emergence of floating consciousness, *Table 6.4* delineated some of the major challenges for the emergence of hyper-spatial consciousness, and *Table 6.5* identified some of the theoretical bases of the existence of different universes, as the three tables were originally worked out and analyzed in *FCD* and, especially, *FPHC*.

The Future of Post-Human Space-Time Unto Multiverses

Whether or not these post-humans will last, or how long they will do so, is not something that current humans can predict with exactitude.

But an educated prediction in my previous works, especially in *FCD* and *FPHC*, is that a final episode of the long evolution of intelligent life is the emergence of what I originally called *floating consciousness* (as summarized in *Table 1.4* here), which, in one of its most expansionist forms, may well be the expansion, in the universe and beyond, of a cosmic consciousness, competing for sure with alternative rivals.

And in *BCIV*, I further predicted how intelligent life in its current civilizational forms will eventually evolve into its "post-civilizational" counterparts (as summarized in *Table 1.30*, *Table 1.31*, *Table 1.32*, and *Table 1.33* here), especially though not exclusively in the post-human era.

In *FCD* (480-1), I thus wrote about the future of intelligent life in the cosmos and beyond: "And if gazed at the farther stretch of spacetime, the rivalry for cosmic hegemony among competing floating lifeforms (or whatever else totally unknown yet of future post-humans, or even extra-terrestrial beings)...will perhaps render the very idea of the highest state of freedom at the rational endpoint of history all the more wishful-thinking, though consoling to the pitiful human mind of the present. Should one of these competing cosmic hegemons one day succeed in subduing all others and become the sole remaining one in the cosmos to rule eternally, the history of intelligent life, if carried to that farthest reach in the cosmos, might well end up having its last chapter on a cosmic Dr.Jekkle and Mr.Hyde (with both noble and devilish traits at once)."

The cyclical progression of hegemony, even if expanding on that cosmic scale, is equally ambivalent in its nature, as there is no utopia without distopia in the end. Should there be a "god" in the end, this would well be a most plausible form of its expression, but in a most ambivalent way which shatters into pieces the romantic vision by humans in all history hitherto existing, with all its different forms for the fragile solace of the pitiful human mind.

Even the post-human consciousness itself, be it in the form of floating consciousness, of hyper-spatial consciousness, or of whatever else, and with its myriad versions yet to be developed in most distant history, may as well have its day numbered in the most remote future of space-time—just as no version of space-time and matter-energy will remain unchanged or last forever....

Table 6.1. Types of Super Civilization in the Cosmos (Part I)

• **Type I**
—a civilization which gains control of and uses the total energy output "falling on its planet from its sun for interstellar communication" (or, in general, space colonization). For N.Kardashev, who proposed the first three types, human civilization is currently Type Zero (Type 0), which is below even Type I, since its present energy consumption for all purposes, let alone for interstellar communication, is still 10,000 times less.

• **Type II**
—a civilization which gains control of and uses directly the total energy output of its sun for interstellar communication (or, in general, space colonization).

• **Type III**
—a civilization which gains control of and uses the total energy output of its galaxy for interstellar communication (or, in general, space colonization).

• **Type IV**
—a civilization which gains control of and uses the total energy output of its cluster of galaxies for interstellar communication (or, in general, space colonization).

• **Type V**
—a civilization which gains control of and uses the total energy output of its supercluster of galaxies for interstellar communication (or, in general, space colonization).

(continued on next page)

Table 6.1. Types of Super Civilization in the Cosmos (Part II)

- **Type...n**
—So continues the series in what I call *the cyclical progression of hegemony* in the cosmos and beyond.

Notes: The Russian astrophysicist Nikolai Kardashev proposed the first three types of super civilization in terms of total energy output for interstellar communication.(CSM 1979) I extend his argument further to propose Type IV, Type V, Type VI, and Type...n, in the context of my claim about the cyclical progression of hegemony in the cosmos and beyond.

Sources: From *Table 9.4* in *FCD*. See *FHC*, *FCD*, and *FPHC* for more info.

Table 6.2. The Technological Frontiers of the Micro-World

- **Type I-Minus**
  —Ex: building structures and mining

- **Type II-Minus**
  —Ex: playing with the genetic makeups of living things

- **Type III-Minus**
  —Ex: manipulating molecular bonds for new materials

- **Type IV-Minus**
  —Ex: creating nanotechnologies on the atomic scale

- **Type V-Minus**
  —Ex: engineering the atomic nucleus

- **Type VI-Minus**
  —Ex: restructuring most elementary particles

- **Type Ω-Minus**
  —Ex: altering the structure of space-time

Notes: As already indicated in *Sec.4.4.2.2* of *FPHC*, the problem with this micro-classification (from Barrow's work) is that the civilization types (with the exception of Type Ω-Minus, for example) are not quite distinct, since many of them can be achieved more or less in a civilization, to the extent that Type II-minus and Type III-minus, just to cite two plausible types, can be historically contemporaneous, relatively speaking, unlike the vast historical distance between, say, Type 0 and Type I (or Type I and Type II) civilizations. In other words, the micro-classification here is not very useful to understand civilization types but is revealing to see the technological frontiers of the micro-world.

Sources: A reconstruction from J.Barrow (1998:133), as originally shown in *Table 4.7* in *FPHC*. See *FCD* and *FPHC* for more info.

Table 6.3. Main Reasons for Altering Space-Time

- **The Need to Make New Energy-Matter**
  - Ex: manipulating molecular bonds for new materials
  - Ex: creating nanotechnologies on the atomic scale
  - Ex: engineering the atomic nucleus
  - Ex: restructuring most elementary particles
  - Ex: inventing new forms of matter and energy

- **The Need to Create New Space-Time**
  - Ex: creating `warp drive´ (as in science fiction) for space travel
  - Ex: creating "pocket universes"

- **The Need to Conquer the Cosmos unto Multiverses**
  - Ex: spreading floating consciousness and hyper-spatial consciousness, besides other forms that humans have never known, in the cosmos and beyond unto multiverses for ultimate conquest

Notes: The examples in each category are solely illustrative (not exhaustive) nor necessarily mutually exclusive, and the comparison is relative (not absolute). As generalities, they allow exceptions. Also, it should be stressed that the three reasons are all related, in that they all contribute to the evolution of intelligent life in the cosmos unto multiverses in the most distant future beyond our current knowledge.

Sources: A summary of *Sec.6.2*. See also *FHC*, *FCD*, and *FPHC*.

Table 6.4. Physical Challenges to Hyper-Spatial Consciousness

- **The Understanding of a Higher-Dimensional World of Space-Time**
  - Ex: 4 for traditional aspects of space-time (e.g., length, width, breadth and time) plus 6 more new dimensions in theory of hyper-space, with profound implications for practical applications to new forms of consciousness

- **The Mastering of Dark Matter and Dark Energy**
  - Ex: "ordinary matter" (e.g., atoms, molecules) as a mere 4.4% of the universe, with 23% made of "cold dark matter" and the rest (about 73%) of mysterious "dark energy," with fundamental significance to questions about the limit of the speed of energy (or info), the availability of energy for use, and the nature of space-time, just to cite some examples

- **The Exploration of Multiverses**
  - Ex: theoretical speculation of other universes (e.g., "baby universes," "gateways" in black holes, "wave function of the universe," "many worlds," "brane worlds"), with potentially seminary discoveries of different physical laws in relation to matter-energy and space-time, and vital differences to the future of post-human conquest of other universes (for the emergence of new forms of consciousness)

Notes: These examples are solely illustrative (not exhaustive), and some of the items can be reclassified somewhere else. Nor are they always mutually exclusive. Since they are generalities, exceptions are expected. The point here is to give a rough picture of the evolution of consciousness to the hyper-spatial consciousness and others totally unknown to current earthlings. As a note of clarification, it makes no difference to my argument as to whether or not the hyper-spatial consciousness may emerge before, during, and after floating consciousness.
Sources: From *Table 4.5* of *FPHC*. See *FHC*, *FCD* and *FPHC* for more info.

Table 6.5. Theoretical Speculations of Multiverses

- **"Baby Universes" (Ex: Andre Linde and others)**
  - Ex: In a flat universe theory, "even if our part of it eventually collapses,...some spots in the cosmos would suddenly start inflating on their own, creating brand-new `baby universes.´" (P. Baofu 2000: 623)

- **"Parallel Universes" (Ex: Stephen Hawking and others)**
  - Ex: In quantum cosmology, there allows the existence of infinite numbers of parallel universes, with tunneling among them. (M. Kaku 1994: 256) Hawking later revised his views on this.

- **"Pocket Universes" (Ex: Alan Guth)**
  - Ex: "As the pocket universes live out their lives and recollapse or dwindle away, new universes are generated to take their place....While life in our pocket universe will presumably die out, life in the universe as a whole will thrive for eternity." (A. Guth 1997: 248; P. Baofu 2002: 482)

- **"Brane Worlds" (Ex: Warren Siegel, Lisa Randall, and others)**
  - Ex: Our universe is stuck on a membrane of space-time embedded in a larger cosmos, with different brane worlds connecting and/or colliding with each other.

Notes: These examples are solely illustrative (not exhaustive), and some of the items can be reclassified somewhere else. Nor are they always mutually exclusive. Since they are generalities, exceptions are expected.

Sources: From *Table 4.8* of *FPHC*. See *FHC*, *FCD* and *FPHC* for more info.

Bibliography

Arisaka, Yoko. 1996. "Spatiality, Temporality, and the Problem of Foundation in Being and Time." *Philosophy Today* 40: 1 (Spring): 36-46. <http://www.arisaka.org/heidegger.html>.

Baofu, Peter. 2006. *Beyond Civilization to Post-Civilization: Conceiving a Better Model of Life Settlement to Supersede Civilization.* New York: Peter Lang Publishing, Inc.

_____.2005. *Beyond Capitalism to Post-Capitalism: Conceiving a Better Model of Wealth Acquisition to Supersede Capitalism.* New York: The Edwin Mellen Press.

_____.2004. *The Future of Post-Human Consciousness.* New York: The Edwin Mellen Press.

_____.2004a. *Beyond Democracy to Post-Democracy: Conceiving a Better Model of Governance to Supersede Democracy.* 2 volumes. New York: The Edwin Mellen Press.

_____.2002. *The Future of Capitalism and Democracy.* Maryland: The University Press of America.

_____.2000. *The Future of Human Civilization.* 2 volumes. New York: The Edwin Mellen Press.

Baard, Mark. 2006. "MIT Interns Prepare for an Economy That Looks to Space" (July 3). *The Boston Globe.* <http://www.boston.com/business/technology/articles/2006/07/03/mit_interns_prepare_for_an_economy_that_looks_to_space/>.

Barrow, John D. 1998. *Impossibility: The Limits of Science and the Science of Limits.* Oxford: Oxford University Press.

Bartleby.com (BT). 2005. "William James." <http://www.bartleby.com/73/1837.html>.

_____.2005a. "Jeremy Hayward." <http://www.bartleby.com/66/97/27397.html>.

_____.2005b. "Herbert George Wells." <http://www.bartleby.com/66/96/63696.html>.

BBC.co.uk (BBC). 2005. "Parallel Universes." <http://www.bbc.co.uk/science/horizon/2001/paralleluni.shtml>.

Beldona, Sam, Andrew Inkpen & Arvind Phatak. 1998. "Are Japanese Mangers More Long-Term Oriented Than United States Managers?" *Management International Review*. Weisbaden: Third Quarter.

Berg, Berg, Eileen Applebaum, Tom Bailey & Arne Kalleberg. 2004. "Contesting Time: International Comparisons of Employee Control of Working Time." *Industrial and Labor Relations Review*, vol.57, no.3 (April): 331-349.

Bond, Michael & Norman Feather. 1988. "Some Correlates of Structure and Purpose in the Use of Time." *Journal of Personality and Social Psychology*, 55 (August): 321-29.

Brislin, Richard & Eugene Kim. 2003. "Cultural Diversity in People's Understanding and Uses of Time." *Applied Psychology: An International Review*, 52(3): 363-382.

Brown, Reva Berman & Richard Herring. 1998. "The Circles of Time: An Exploratory Study in Measuring Temporal Perceptions Within Organizations." *Journal of Managerial Psychology*. United Kingdom: Bradford.

Calabresi, Renata & Jacob Cohen. 1968. "Persoanlity and Time Attitudes." *Journal of Abnormal Psychology*, 73 (5): 431-39.

Carnevale, Anthony. 1991. *America and the New Economy*. VA: American Society for Training and Development.

CNN.com (CNN). 2004. "U.S. Scientists Create New Form of Matter: Finding CouldLead to Better Superconductors" (Thursday, January 29). <http://edition.cnn.com/2004/TECH/science/01/28/matter.new.reut/index.html>.

Conway, Lucian. 2004. "Social Contagion of Time Perception." *Journal of Experimental Social Psychology*, 40: 113-120.

Cosmic Search Magazine (CSM). 1979. "Glossary." Cosmic Quest, Inc. <http://www.bigear.org/vol1no1/glossary.htm>.

Cotte, June, S. Ratnershwar & David Mick.. 2004. "The Times of Their Lives: Phenomenological and Metaphorical Characteristics of Consumer Timestyles." *Journal of Consumer Research* (September): 333-345.

Cottle, Thomas & Stephen Klineberg. 1974. *The Present of Things Future: Explorations of Time in Human Experience*. New York: The Free Press.

Davies, Paul. *God and the New Physics*. NY: Simon & Schuster, 1983.

Discovery Wings Channel (DWC). 2001. "Destination Space." Cable Channel 112 (Tuesday, October 30).

Durkee, N. & R. Kastenbaum. 1964. "Elderly People View Old Age." *New Thoughts on Old Age*. Ed., R.Kastenbaum. New York: Springer-Verlag.

Fast, Julius. 1970. *Body Language*. New York: M. Evans and Co.

Fine, C.H. 1998. *Clockspeed: Winning Industry Control in the Age of Temporary Advantages*. Reading, MA: Perseus Books.

Fraisse, P. 1964. *The Psychology of Time*. London: Eyre & Spottiswoode Ltd.
Giedion, Sigfried. 1976. *Space, Time and Architecture*. Cambridge, MA: Harvard University Press.
Givens, David. 2003. "Proxemics." Center for Non-Verbal Studies. <http://members.aol.com/doder1/proxemi1.htm>.
Goulet, Laurel. 1999. "How Managers Control Employees' Time." *The Academy of Management Executive* (August): 114-115.
Grenier, Guillermo. 2002. "Types of Human Societies." <http://www.fiu.edu/~grenierg/chapter4.htm>.
Gribbin, John & Mary Gribbin. 1994. *Time and Space*. New York: Dorling Kindersley.
Guth, Alan. 1997. *The Inflationary Universe: The Quest for a New Theory Of Cosmic Origins*. New York: Helix Books.
Halber, Deborah. 2006. "Physicists Get to Heart of Antimatter" (April 12). *MIT News*. <http://web.mit.edu/newsoffice/2006/antimatter-0412.html>.
Hall, Edward. 1969. *The Dance of Life: The Other Dimension of Time*. New York: A Division of Random House, Inc.
Hancock, P.A. 1997. "On the Future of Work: How the Future Promises To Alter Our Concept of Work." *Ergonomics in Design* (October): 25-29.
Harvey, Michael & Milorad Novicevic. 2001. "The Impact of Hypercompetitive `Timescapes´ on the Development of a Global Mindset." *Management Decision*: 448-460.
Heidegger, Martin. 1962. *Being and Time*. E. Robinson and J. Macquarrie, trs. New York: Harper.
Idea Works, Inc. (IW). 1995. "Sociology Timeline." <http://www.missouri.edu/~socbrent/timeline.htm>.
Institute for Shipboard Education (ISE). 2005. *Port-to-Port Global Studies*. NY: McGraw Hill.
Irani, George. 2000. "Rituals of Reconciliation: Arab-Islamic Perspectives." *Kroc Institute Occasional Paper* #19:OP:2.
Kaku, Michio. 1994. *Hyperspace: A Scientific Odyssey Through Parallel Universes, Time Warps, and The Tenth Dimension*. Illustrations by Robert O'Keefe. Oxford: Oxford University Press.
Karasek, R.A. 1979. "Job Demands, Job Decision Latitude and Mental Strain: Implications for Job Redesign." *Administrative Science Quarterly*, 24: 285-308.
Kastenbaum, R. 1975. "Time, Death and Ritual in Old Age." *The Study of Time II, Proceedings of the Second Conference of the International Society for the Study of Time*. J.Fraser and N.Lawrence, eds. New York: Springer-Verlag.
_____.1964. "Cognitive and Personal Futurity in Later Life." *Journal of Individual Psychology*, vol.19: 216-219.
Lemke, Jay. 2000. "Across the Scales of Time: Artifacts, Activities, and Meanings in Ecosocial Systems." *Mind, Culture and Activity*, 7(4): 273-290.

_____.2000a. "Material Sign Processes and Ecosocial Organization." *Downward Causation: Self-Organization in Biology, Psychology, and Society.* P.B.Andersen, C.Emmeche and N.O.Finnemann-Nielsen, eds. Aarhus, Denmark: Aarhrus University Press.

Marx, Karl. 1999. *Capital.* Volume 3. Transcribed by M. Griffin and others. <http://www.marxists.org/archive/marx/works/1894-c3/ch48.htm>.

Massachusetts Institute of Technology News Office (MIT). 2006. "MIT Physicists Shed New Light on Superfluidity" (July 20). <http://web.mit.edu/newsoffice/2006/superfluidity.html>.

McAleer, Neil. 1987. *The Mind-Boggling Universe.* New York: Doubleday & Co.

Melges, Frederick. 1982. *Time and the Inner Future.* New York: Wiley.

Merriam-Webster's Collegiate Dictionary (MWCD). 2003. "Space." Springfield, MA: Merriam-Webster.

_____2003a. "Space." Springfield, MA: Merriam-Webster.

Merriam-Webster's Collegiate Dictionary Online (MWD). 2005. "Space." <http://www.m-w.com/cgibin/dictionary?book=Dictionary&va=space> (January 29).

_____.2005a. "Time." <http://www.m-w.com/cgi-bin/dictionary?book=Dictionary &va=time> (January 29).

_____.2005b. "Culture." <http://www.m-w.com/cgi-bin/dictionary?book=Dictionary&va=culture> (January 31).

_____.2005c. "Society." <http://www.m-w.com/cgi-bin/dictionary?book=Dictionary&va=society> (February 01).

_____.2005d. "Mind." <http://www.m-w.com/cgi-bin/dictionary?book=Dictionary&va=mind> (February 02).

_____.2005e. "Nature." <http://www.m-w.com/cgi-bin/dictionary?book=Dictionary&va=nature> (February 02).

Msumange, G. 1998. "Time – The Concept of Future in the Thought of the `Hebe´ and Other African Tribes." Unpublished Manuscript. S. Major Seminary. Morogoro, Tanzania.

Murphy, Caryle. 2001. "Bin Laden's Radical Form of Islam." *The Washington Post* (September 18): A.23.

NewScientist (NS). 2004. "Civilization: How Human Nature Created the Modern World" (special Issue, September 18-24): 26.

Orlikowski, Wanda. 2002. "It's About Time: Temporal Structuring in Organizations." *Organizational Science*, vol.13, no.6, November-December: 684-700.

Pearson, Fredric & Simon Payaslian. 1999. *International Political Economy: Conflict and Cooperation in the Global System.* New York: McGraw-Hill.

People's Daily Online (PD). 2006. "Researchers Discover Another Particle" (February 07). <http://english.peopledaily.com.cn/200602/07/eng20060207_240784.html>.

_____.2005. "China Tests Secure Quantum Communication."

<http://english.peopledaily.com.cn/200511/29/eng20051129_224358.html>.

Pickover, Clifford. 1998. *Time: A Traveler's Guide*. New York: Oxford University Press.

Poole, Barbara. 2000. "On Time: Contributions from the Social Sciences." *Financial Services Review*, 9: 375-387.

Pravda.ru (PR). 2006. "Spaceships of the Future to Take Humans to Mars in 2.5 Hours" (February 16). <http://english.pravda.ru/science/tech/16-02-2006/76045-0>.

Saunders, Carol. 2005. "International Dimensions of Organizational Behavior." University of Pittsburgh: Spring semester-at-sea.

Schriber, Jacqueline & Barbara Gutek. 1987. "Some Dimensions of Work: Measurement of an Underlying Aspect of Organization Culture." *Journal of Applied Psychology*, 72: 642-650.

SCICH. 2002. "Cosmic Safari." TV Cable Channel 110 (Saturday, April 13).

Servan-Schrieber, J.L. 1988. *The Art of Time*. New York: Addison-Wesley.

Shlain, Leonard. 1991. *Art and Physics: Parallel Visions in Space, Time, and Light*. New York: William Morrow and Co., Inc.

Sklar, Lawrence. 1974. *Space, Time, and Spacetime*. CA: University of California Press.

Stix, Gary. 2002. "Real Time." *Scientific American* (September): 36-39.

Stuttman, Inc (Stuttman). 1992. *Mysteries of Mind, Space and Time: The Unexplained*. Westport, Connecticut: H.S.Stuttman, Inc.

Time-Life Books Inc. (Time). 1990. *Time and Space*. Richmond, Virginia: The Time-Life Books Inc.

Valigra, Lori. 2005. "MIT Physicists Create New Form of Matter." *MIT News*. (June 22). <http://web.mit.edu/newsoffice/2005/matter.html>.

Van Slyke, Craig, Douglas Vogel & Carol Saunders. 2004. "My Time or Yours? Managing Time Visions in Global Virtual Teams." *Academy of Management Executive*, vol.18, no.1: 19-31.

Vinton, Donna. 1992. "A New Look at Time, Speed, and the Manager." *Academy of Management Executive*, vol.6, no.4: 7-16.

Wikipedia (WK). 2004. "Feudalism." <http://en.wikipedia.org/wiki/Feudalism>.

_____.2004a. "Neo-Classical Economics." <http://en.wikipedia.org/wiki/Neoclassical_economics>.

_____.2004b. "Monetarism." <http://en.wikipedia.org/wiki/Monetarism>.

_____.2004c. "New Classical Economics." <http://en.wikipedia.org/wiki/New_classical_economics>.

_____.2004d. "Rational Expectations." <http://en.wikipedia.org/wiki/Rational_expectations>.

_____.2004e. "Physiocrats." <http://en.wikipedia.org/wiki/Physiocrat>.

_____.2004f. "Francois Quesnay." <http://en.wikipedia.org/wiki/Fran%E7ois_Quesnay>.

Wyrick, R. & L.Wyrick. 1977. "Time Experience During Depression." *Archives of General Psychiatry*, 34: 1441-1443.

Index

•A•

Aalto, Alvar, 98
absolutist perspective of space-time,
      3–5, 7–8, 18–19
      see also Space, space-time, time
aesthetic space, 101, 108
      see also Space
Africa
      and linear vs. cyclic time,
           103–106
after-postmdernism
      see After-Postmodernity
after-postmodernity, 8–14
      and its trinity, 12, 69
      from pre-modernity to after-
           postmodernity, 70–71
      see also Existential dialectics
after-postmdernization
      see After-Postmodernity
age, 121–122
      see also Discriminatory space
agents
      from pre-modernity to after-
           postmodernity, 70–71
Alculierre, Miguel, 170
animals
      in relation to time, 137–144
      see also Time
anti-gravitation, 169–170

Applebaum, Eileen, 120
architecture
      and space, 96–101, 124–129
Arisaka, Yoko, 99
art
      in relation to space, 101, 108,
           155–156
asymmetry
      and symmetry, 12–3
      and the symmetry-asymmetry
           principle, 13, 19
      see also Existential dialectics
atomic particles
      and the technological conquest
           of the micro-world, 175
autonomy space
      see Space

•B•

baby univeres
      in relation to the physical
           challenges to hyper-spatial
           consciousness, 177
      in relation to the theoretical
           speculations of multiverses,
           178
Bailey, Tom, 120
Baofu, Peter, xiii

and his idea of the foundation
fallacy in the theoretical debate
on space-time, 8, 19
and his perspectival theory of
space-time, 7–8, 18–19
(1)multiple perspective of
space-time, 7–8, 19
(2)some perspectives more
hegemonic, 8, 19
(3)post-human forms of
space-time, 8, 19
and his theories on civilizational
holism, 86–91
barbarity
see Civilization, existential
dialectics
Baroque, the, 124–126
see Architecture, space
beautiful, the
and existential dialectics, 9
and the equality/inequality, 9,
24
and the freedom/unfreedom
dialectics, 21
in relation to
moderntiy, 65–66
postmodernity, 67–68
behavior
in relation to time, 143–144
see also Time
being
and the equality/inequality, 9,
24
and the freedom/unfreedom
dialectics, 20–22
belonging
and the equality/inequality, 9,
23
and the freedom/unfreedom
dialectics, 20–22
Bennett, M., 119
Berg, Peter, 120
Big Bang, the, 154–156, 165–169
see also Space
biological, the
and civilizational holism, 86–91

and floating consciousness, 25
and space-time, 138–140, 146
and the multiple causes of the
emergence of post-capitalism,
48–9
and the multiple causes of the
emergence of post-democracy,
58–59
see Methodological holism
biospace
see Space
biotime
see Time
black hole, 154–158, 164
in relation to the physical
challenges to hyper-spatial
consciousness, 177
see Multiverse, universe
body functions
in relation to time, 139–140
see also Time
Bose-Einstein condensate
see Matter
brane world
in relation to the physical
challenges to hyper-spatial
consciousness, 177
in relation to the theoretical
speculations of multiverses,
178
Brown, Reva Berman, 114, 123

•C•

capitalism
and after-postmodernity, 69
and modernity, 65–66
and postmodernity, 67–68
and pre-modernity, 63–4
from pre-modernity to after-
postmodernity, 70–71
in relation to non-capitalism,
and post-capitalism, 44–47
see also Post-capitalism
capitalist value ideals, 29–33

classical, 31
Keynesian, 32
monetarist, 32
neo-classical, 31
neo-mercantilist, 33
new classical, 33
see also Value ideals
Carnevale, Anthony, 121
challenge
   and the post-human alteration of space-time, 165–171, 176
   and the post-human encounter, 164
   and the post-human future, 171–172
   see Multiverse, space-time
change
   and change-constancy principle, 13, 19, 81–85
   see also Existential dialectics
chemical, the
   and civilizational holism, 86–91
   and floating consciousness, 25
   and space-time, 136–138, 146
   and the multiple causes of the emergence of post-capitalism, 48–9
   and the multiple causes of the emergence of post-democracy, 58–59
   see Methodological holism
chemospace
   see Space
chemotime
   see Time
Chinese, the
   and their understanding of space, 113, 139
   and their understanding of time, 105
civilization
   and the theses on post-civilization, 77
   and types of super-civilization in the cosmos, 173–174
   the four processes of, 76, 78–81
   from pre-modernity to after-postmodernity, 70–71
   see Existential dialectics
civilization/barbarity dialectics
   see Existential dialectics
civilizational holism, 86–91
   at different levels, 86–88
   theories on, 89–91
class, 121–122
   see also Discriminatory space
classical economics, 31
classical mechanics, 3–5
cognition
   in relation to time, 143–144
   see also Time
colonization
   see Hegemony
communication
   in relation to space, 141–144, 151
   see also Space
commumal
   and capitalist economics, 29–33
   and eco-feminist economics, 34
   and feudalist economics, 27
   and hunting/gathering economics, 27
   and Islamic economics, 35
   and Marxist economics, 34
   and mercantilist economics, 28
   and physiocratic economics, 28
   and post-capitalism, 36–43
   see also Post-human elitist calling, trans-feminine calling, trans-Islamic calling, trans-outerspace calling, trans-Sinitic calling
competitive space
   see Space
competitive time
   see Time
conscious
   in relation to time, 142–143
   see also Time
constancy
   and the change-constancy principle, 13, 19

see also Existential dialectics
continuous time
   see Time
cooperative space
   see Space
cooperative time
   see Time
coordination time, 115–116, 132
   see also Time
cosmological, the
   and civilizational holism, 86–91
   and floating consciousness, 26
   and the East, 155–156
   and the multiple causes of the emergence of post-capitalism, 48–9
   and the multiple causes of the emergence of post-democracy, 58–59
   and the post-human alteration of space-time, 165–171, 176
   and the post-human challenge, 164
   and the post-human future, 171–172
   and types of super-civilization in the cosmos, 173–174
   in relation to space, 154–156
   in relation to time, 156–158
   see Methodological holism
Cottle, Thomas, 123–124
Cubism, 98–99
culture
   defined, 95–96
   and methodological holism, 14–15
   and the civilizing process, 76, 78–81
   in relation to space and time, 95–108
cultural, the
   and civilizational holism, 86–91
   and floating consciousness, 25
   and the multiple causes of the emergence of post-capitalism, 48–9
   and the multiple causes of the emergence of post-democracy, 58–59
   in relation to space and time, 95–108
   see Methodological holism
cyborgs, 9–14
   see also Post-humans
cylcial progression of hegemony, 12, 69
   and types of super-civilization in the cosmos, 173–174
   from pre-modernity to after-postmodernity, 70–71
   see also Existential dialectics, hegemony
cyclic time, 102–103, 108
   see also Time

•D•

Daly, Cesar, 125
dark energy
   in relation to the physical challenges to hyper-spatial consciousness, 177
   see also Space-time
dark matter
   in relation to the physical challenges to hyper-spatial consciousness, 177
   see also Space-time
Dealy, Glen, 103
d' Eaubonne, Francois, 34
democracy
   from pre-modernity to after-postmodernity, 70–71
   in relation to non-democracy, and post-democracy, 55–57
   see also Post-democracy
depth
   and dimensions of space-time, 4
   see also Space, space-time
dimension
   and space-time, 4

see also Space, space-time
directionality, 99
   see also Space
discriminatory space
   see Space
discriminatory time
   see Time
distance
   in relation to space, 141–144
   see also Space
distance space
   see Far space
Dunne, John William, 158
Durkee, N., 123
Dyson, Freeman, 165

• E •

East, the
   in relation to space, 113, 139, 155–156
ecobiotime
   see Time
eco-feminist economics
   and alternatives to capitalist value ideals, 33
economic revolutions
   from pre-modernity to after-postmodernity, 70–71
economics
   see Value ideals
ecosocial space
   see Space
ecosocial time
   see Time
Einstein, Albert, 5–7, 18–19, 157
   see also Space, space-time, time
Ekelof, Tord, 168
emotions
   in relation to time, 142–144
   see also Time
empty space
   see Space
energy and matter, 5–7

and the post-human alteration, 165–171, 176
and the post-human challenge, 164
and the post-human future, 171–172
in relation to the physical challenges to hyper-spatial consciousness, 177
theoretical debate on, 3–8
see also Space, space-time, time
Engels, Friedrick, 34
environmental space
   see Space
epistemic space, 96–99, 108
   see also Space
equality/inequality dialectics
   see Existential dialectics
Euclidean geometry, 3–7
   see also Space, space-time, time
environmental space
   see Space
ethnicity, 121–122
   see also Discriminatory space
everyday, the
   and existential dialectics, 9
   and the equality/inequality, 9, 23
   and the freedom/unfreedom dialectics, 20
   in relation to
     moderntiy, 65–66
     postmodernity, 67–68
existential dialectics, 8–14
   and civilization/barbarity, 9, 75
   and earlier works, 8–14
   and equality/inequality, 9, 23–24, 53, 72–73, 78–81
   and freedom/unfreedom, 9, 20–22, 53, 72–73, 78–81
   and its five features, 12–3
   and its ontological logic, 13, 82–85
   and its seven dimensions of existence, 9
   and post-capitalism, 36–49, 60–

62
 and post-democracy, 50–59, 60–62
 and the four processes of, 76, 78–81
 and wealth/poverty, 74
 from pre-modernity to after-postmodernity, 70–71
 no to be confused with existentialism, 9

•F•

fallacy
 see Foundation fallacy
far and near space, 99, 108
 see also Space
Fast, Julius, 141–142
fermionic condensate
 see Matter
feudalist economies, 27
floating consciousness, 9–14, 171, 176
 theory of, 25–6
 see also Post-humans
foundation fallacy, the
 in the theoretical debate on space-time, 8, 19
 see also Space-Time
formal-legalistic
 and capitalist economics, 29–33
 and eco-feminist economics, 34
 and feudalist economics, 27
 and hunting/gathering economics, 27
 and Islamic economics, 35
 and Marxist economics, 34
 and mercantilist economics, 28
 and physiocratic economics, 28
 and post-capitalism, 36–43
 see also Post-human elitist calling, trans-feminine calling, trans-Islamic calling, trans-outerspace calling, trans-Sinitic calling
Fraisse, P., 123

freedom
 and after-postmodernity, 69
 and modernity, 65–66
 and postmodernity, 67–68
 and pre-modernity, 63–4
freedom/unfreedom dialectics
 see Existential dialectics
Friedman, Milton, 32
functionalism, 121
 see also Cooperative space
future
 and linear vs. cyclic time, 103–106
 see also Time

•G•

galaxy, 10
 and types of super-civilization in the cosmos, 173–174
 see also Post-humans
gas
 see Matter
gathering economics, 27
gender, 121–122
 see also Discriminatory space
genetically altered superior beings, 9–14
 see also Post-humans
Germany
 and neo-Mercantilism, 33
Giedion, Sigfried, 97–98, 100–101, 117, 124–126, 128–129
Gleick, James, 106
global, 10
 see also Post-humans
god
 and the future of post-human space-time, 172
good, the
 and existential dialectics, 9
 and the equality/inequality, 9, 23
 and the freedom/unfreedom dialectics, 21

in relation to
    moderntiy, 65–66
    postmodernity, 67–68
Gothic, the, 124–126
    see Architecture, space
Goulet, Laurel, 116
Great Depression, the
    see Capitalism, modernity
Gribbin, John, 138–139
Guth, Alan, 166–167, 169, 178

• **H** •

Hall, Edward, 102–103, 105
Hancock, P.A., 117–118
Harvey, Michael, 129
having
    and the equality/inequality, 9, 23
    and the freedom/unfreedom dialectics, 20–22
Hawking, Stephen, 154, 178
Hayward, Jeremy, 149
hegemony
    and after-postmodernity, 69
    and modernity, 65–66
    and postmodernity, 67–68
    and pre-modernity, 63–4
    cyclical progression of, the, 12
    see also Existential dialectics
Heidegger, Martin, 99, 105
Herring, Richard, 114, 123
Hindu, the
    and their understanding of time, 103–106
Hoagland, Hudson, 140
Hofstede, Geerte, 114
Hoizer, Joachim, 169
holism
    see Methodological holism
Hopi tribe, the
    and their understanding of time, 102–106
holy, the
    and existential dialectics, 9

    and the equality/inequality, 9, 24
    and the freedom/unfreedom dialectics, 21
    in relation to
        moderntiy, 65–66
        postmodernity, 67–68
horizontal space, 98, 108
    see also Space
Hoyle, Fred, 158
hunting/gathering economics, 27
hyper-spatial consciousness, 9–14, 171, 176
    physical challenges to, 177
    see also Post-humans

• **I** •

imaginary time
    see Time
impacts
    from pre-modernity to after-postmodernity, 70–71
imperialism
    see Hegemony
individualistic
    and capitalist economics, 29–33
    and eco-feminist economics, 34
    and feudalist economics, 27
    and hunting/gathering economics, 27
    and Islamic economics, 35
    and Marxist economics, 34
    and mercantilist economics, 28
    and physiocratic economics, 28
    and post-capitalism, 36–43
    see also Post-human elitist calling, trans-feminine calling, trans-Islamic calling, trans-outerspace calling, trans-Sinitic calling
Industrial Revolution, the
    see Capitalism, modernity
inequality
    see Existential dialectics

informal-legalistic
    and capitalist economics, 29–33
    and eco-feminist economics, 34
    and feudalist economics, 27
    and hunting/gathering economics, 27
    and Islamic economics, 35
    and Marxist economics, 34
    and mercantilist economics, 28
    and physiocratic economics, 28
    and post-capitalism, 36–43
    see also Post-human elitist calling, trans-feminine calling, trans-Islamic calling, trans-outerspace calling, trans-Sinitic calling
inner and outer space, 97–98, 108
    see also Space
institutional, the
    and civilizational holism, 86–91
    and floating consciousness, 25
    and the multiple causes of the emergence of post-capitalism, 48–9
    and the multiple causes of the emergence of post-democracy, 58–59
    in relation to space-time, 116–120, 132
    see Methodological holism
instutitions
    from pre-modernity to after-postmodernity, 70–71
international organizations, 121–122
    see also Discriminatory space
interpersonal communication
    and types of super-civilization in the cosmos, 173–174
    in relation to space, 141–144
    see also Space
interstellar space travel, 169–170
    and types of super-civilization in the cosmos, 173–174
Islam
    and moral space, 100
Islamic economics
    and alternatives to capitalist value ideals, 33

•J•

James, Williams, 135
Japan
    and neo-Mercantilism, 33
Jevons, W. Stanley, 31
just, the
    and existential dialectics, 9
    and the equality/inequality, 9, 23
    and the freedom/unfreedom dialectics, 21
    in relation to
        moderntiy, 65–66
        postmodernity, 67–68

•K•

Kalleberg, Arne, 120
Kardashev, Nikolai, 173
Kastenbaum, R., 123
Kelvin, John, 169
Keynes, John M., 32
Keynesian economics
    and capitalist value ideals, 32
Kim, Brislin, 105
Kim, Euguen, 105
Kleitman, Nathaniel, 140
Klineberg, Stephen, 123–124
Kozyrev, Nikolai, 169

•L•

Latin America
    and monochronic vs. polychronic time, 102–103
Lemke, Jay, 127, 130–131
length
    and dimensions of space-time, 4

"spaghettified", 6
  see also Space, space-Time
light
  speed of, and length, 6
linear and cyclic time, 103–106, 108
  see also Time
Linde, Andre, 178
liquid
  see Matter
Lucas, Robert, 33

• **M** •

macro-physical, the
  see Cosmological
map, 121–122
  see also Discriminatory space
Marshall, Alfred, 31
Marx, Karl, 34
Marxian economics
  and alternatives to capitalist
    value ideals, 33
Massachusetts Institute of Technology
  (M.I.T.), 97–98, 167–169–170
matter, 3–7, 167–168
  and its composition in the
    universe, 177
  and the technological conquest
    of the micro-world, 175
  classification of, 167–168
  theoretical debate on, 3–8
  see also Energy, space, space-
    time, time
matter-energy, 5–7
  and the post-human alteration
    of space-time, 165–171, 176
  and the post-human challenge,
    164
  and the post-human future,
    171–172
  and the technological conquest
    of the micro-world, 175
  in relation to the physical
    challenges to hyper-spatial
    consciousness, 177

theoretical debate on, 3–8
  see also Space, space-time, time
McAleer, Neil, 154–155
Meerloo, Joost, 144
methodological holism, 14–15
  cosmological, the, 14–15, 153–158
  cultural, the, 14–15, 95–107
  micro-physical, the, 14–15,
    150–153
  biological, the, 14–15, 138–141
  institutional, the, 14–15, 116–120
  organizational, the, 14–15,
    112–116
  psychological, the, 14–15,
    141–145
  structural, the, 14–15, 120–124
  systemic, the, 14–15, 124–131
micro-physical, the
  and civilizational holism, 86–91
  and floating consciousness, 25
  and the multiple causes of the
    emergence of post-capitalism,
    48–9
  and the multiple causes of the
    emergence of post-democracy,
    58–59
  in relation to space, 150–152
  in relation to time, 152–153
  see Methodological holism
Micro-world, 175
Middle Ages, the, 124–126
  see Architecture, space
Milky Way, the
  see also Space-time
mind, the
  and methodological holism,
    14–15
  and the civilizing process, 76,
    78–81
  in relation to space and time,
    135–146
mixing
  see Matter
modernism
  see Modernity
modernization

see Modernity
modernity, 8–14
   and its trinity, 12, 65–66
   from pre-modernity to after-
      postmodernity, 70–71
   see also Existential dialectics
modernization
   see Modernity
monetarist economics
   and capitalist value ideals, 32
monetary time, 119–120, 132
   see also Time
monochronic and polychronic time,
   102–103, 108
   see also Time
moral space, 100, 108
   see also Space
multi-dimensional space
   see Space
multiverse, 10
   and pocket universe, 166
   and the perspectival theory of
      space-time, 7–8, 19
   and the post-human alteration
      of space-time, 165–171
   and the post-human challenge,
      164
   and the post-human future,
      171–172
   and the theoretical debate on, 171
   and types of super-civilization,
      173–174
   in relation to the physical
      challenges to hyper-spatial
      consciousness, 177
   see also Post-humans
Muth, John, 33

•N•

naked contingency
   from pre-modernity to after-
      postmodernity, 70–71
   see also After-Postmodernity
narratives
   from pre-modernity to after-
      postmodernity, 70–71
nature
   and methodological holism,
      14–15
   and the civilizing process, 76,
      78–81
   in relation to space and time,
      149–159
   see Technological
near space, 99, 108
   see also Space
neo-classical economics
   and capitalist value ideals, 31
neo-Mercantilism
   and capitalist value ideals, 33
neologisms
   as current intellectual
      convenience, 17–8
new classical economics
   and capitalist value ideals, 33
Newton, Isaac, 3–5, 18–19
   see also Space, space-time, time
non-capitalism
   in relation to capitalism,
      and post-capitalism, 44–47
   see also Post-capitalism
non-democracy
   in relation to democracy,
      and post-democracy, 55–57
   see also Post-democracy
Novicevic, Milorad, 129

•O•

oppression
   and self-oppression, 12–3
   see also Existential dialectics
organizational, the
   and civilizational holism, 86–91
   and floating consciousness, 25
   and the multiple causes of the
      emergence of post-capitalism,
      48–9
   and the multiple causes of the

emergence of post-democracy, 58–59
in relation to space-time, 112–116, 132
see Methodological holism
organizational hierarchy, 114
see also Space
organizational management, 113–116
see also Space
Orlikowski, Wanda, 119
Others, the
vs. the Same, 12–3
see also Existential dialectics
outcomes
from pre-modernity to after-postmodernity, 70–71
outer space, 97–98, 108
see also Space
Ovrut, Burt, 166

•P•

pacifying process, the
see Civilization
Pan, Jianwei, 151
parellel universes
in relation to the theoretical speculations of multiverses, 178
particles
and the technological conquest of the micro-world, 175
past
and linear vs. cyclic time, 103–106
see also Time
Perlow, Leslie, 116
personality
in relation to time, 143–144
see also Time
perspectival theory of space-time, the 7–8, 19
and the idea of the foundation fallacy in the theoretical debate on space-time, 8, 19
and the three principles in the ontological logic of existential dialectics, 7–8, 19
as a convenient intellectual term, 17–8
three major theses in,
(1) multiple perspective of space-time, 7–8, 19
(2) some perspectives more hegemonic, 8, 19
(3) post-human forms of space-time, 8, 19
Peters, Arno, 122
physics
see Macro-physics, micro-physics
Pickover, Clifford, 157
plasma
see Matter
Planck, Max, 152
Planck's constant, 152–153
pocket universe, 166, 176
in relation to the theoretical speculations of multiverses, 178
see Space-time, universe
polychronic time, 102–103, 108
see also Time
post-capitalism, 11
and floating consciousness, 25
as a convenient intellectual term, 17–8
clarifications about, 60–62
different forms of,
spiritual/individualistic, 42–43
spiritual/communal, 36–41
from pre-modernity to after-postmodernity, 70–71
in relation to non-capitalism, and capitalism, 44–47
multiple causes of the emergence of, 48–9
see also Post-humans
post-civilization
and existential dialectics, 8–14
as a convenient intellectual term,

17–8
    from pre-modernity to after-
       postmodernity, 70–71
    the theses on, 77
post-democracy, 11
    and floating consciousness, 25
    as a convenient intellectual term,
       17–8
    clarifications about, 60–62
    different forms of,
       equality over freedom, 52–53
       freedom over equality, 50–51
       transcending freedom and
          equality, 53–54
    from pre-modernity to after-
       postmodernity, 70–71
    in relation to non-democracy,
       and democracy, 55–57
    multiple causes of the emergence
       of, 58–59
    see also Post-humans
post-human alteration of space-time,
    165–171
    to conquer the cosmos unto
       multiverses, 170–171
    to create new space-time,
       169–170
    to make new matter-energy,
       165–169
post-human challenge, 164
post-human consciousness, 11
    as a convenient intellectual term,
       17–8
    to conquer the cosmos unto
       multiverses, 170–171
    to create new space-time,
       169–170
    to make new matter-energy,
       165–169
    see also Post-humans
post-human counter-elitists, 11
    as a convenient intellectual term,
       17–8
    see also Post-humans
post-human elitists, 11, 42–3, 46, 56

    as a convenient intellectual term,
       17–8
    see also Post-humans
post-human space-time, 171–172
    see also Space-time
post-humanity, 10–14
    as a convenient intellectual term,
       17–8
    from pre-modernity to after-
       postmodernity, 70–71
    to conquer the cosmos unto
       multiverses, 170–171
    to create new space-time,
       169–170
    to make new matter-energy,
       165–169
    see also Post-humans
post-humans
    and after-postmodernity, 9–14
    and post-capitalism, 36–43
    as a convenient intellectual term,
       17–8
    from pre-modernity to after-
       postmodernity, 70–71
    the genealogical origin of the idea
       of, 164
    to conquer the cosmos unto
       multiverses, 170–171
    to create new space-time,
       169–170
    to make new matter-energy,
       165–169
    see also Existential dialectics
post-modernism
    see Post-modernity
post-modernity, 8–14
    and its trinity, 12, 67–68
    from pre-modernity to after-
       postmodernity, 70–71
    see also Existential dialectics
post-modernization
    see Post-modernity
poverty
    and existential dialectics, 74
power space
    see Space

pre-capitalist value ideals
    see Value ideals
pre-modernism
    see Pre-modernity
pre-modernity, 8–14
    and its trinity, 12, 63–64
    from pre-modernity to after-
        postmodernity, 70–71
    see also Existential dialectics
pre-modernization
    see Pre-modernity
present
    and linear vs. cyclic time,
        103–106
    see also Time
production space
    see Space
progression-regression principle, the,
    13, 19
    see also Existential dialectics
psychological, the
    and civilizational holism, 86–91
    and floating consciousness, 25
    and space-time, 141–144, 146
    and the multiple causes of the
        emergence of post-capitalism,
        48–9
    and the multiple causes of the
        emergence of post-democracy,
        58–59
    see Methodological holism
psychospace
    see Space
psychotime
    see Time

•Q•

quantum cosmology, 178
    see also Cosmological
quantum fluctuations
    and the Big Bang, 165
quantum-mechanical space
    see Space
quantum-mechanical time
    see Time

•R•

race, 121–122
    see also Discriminatory space
Randall, Lisa, 178
rationalizing process, the
    see Civilization
real time, 118–119, 132
    see also Time
region, 121–122
    see also Discriminatory space
regression
    and the progression-regression
        principle, 13, 19, 81–85
    see also Existential dialectics
relational space
    see Space
relationality, 98–99
    see also Space
relativist perspective of space-time,
    5–7, 7–8, 18–19
    see also Space, space-time, time
relativity, theory of, 5–7, 167, 169
Renaissance, the, 124–126
    see Architecture, space
resentment
    see Hegemony
responsibility time, 114–115, 132
    see also Time
reversible time
    see Time
Ricardo, David, 31

•S•

Same, the
    vs. the Others, 12–3
    see also Existential dialectics
Saunders, Carol, 115
Schriber, Jacqueline, 115–116
science

the adolescence of, 7
the childhood of, 7
secular
    and capitalist economics, 29–33
    and eco-feminist economics, 34
    and feudalist economics, 27
    and hunting/gathering
        economics, 27
    and Islamic economics, 35
    and Marxist economics, 34
    and mercantilist economics, 28
    and physiocratic economics, 28
    and post-capitalism, 36–43
    see also Post-human elitist
        calling, trans-feminine calling,
        trans-Islamic calling,
        trans-outerspace calling,
        trans-Sinitic calling
self-oppression
    and oppression, 12–3
    see also Existential dialectics
Shlain, Leonard, 3, 18–19, 101,
    155–156
shortest time
    see Time
shrinking
    and stretching of
        space-time, 7–8, 19, 164, 170
Siegel, Warren, 178
simultaneous and successive time,
    106–106, 108
    see also Time
singularity point
    in relation to time, 156–158
Sklar, Lawrence, 157–158
smallest space
    see Space
Smith, Adam, 31
social institutions
    see Space, time
social organizations
    see Space, time
social structure
    see Space, time
social systems
    see Space, time

society
    and methodological holism,
        14–15
    and the civilizing process, 76,
        78–81
    in relation to space and time,
        112–132
solar system , 10
    see also Post-humans
solid
    see Matter
space, 3–8
    and culture, 95–101, 108
        aesthetic space, 101, 108
        epistemic space, 96–99, 108
        moral space, 100, 108
    and nature, 149–160
        in macro-physics, 154–156,
            160
            empty space, 155–56
            multi-dimensional space,
                154, 160
        in micro-physics, 150–152, 160
            quantum-mechanical space,
                151–152, 160
            smallest space, 152, 160
    and society, 111–132
        in social organizations,
            113–108
            power space, 114
            relational space, 113
        in social institutions,
            117–118
            autonomy space, 117–118
            production space, 117
        in social structure,
            120–121
            cooperative/competitive
                space, 114
            discriminatory space,
                121–122
        in social systems,
            124–128
            ecosocial space, 127–128
            environmental space,
                126–127

· INDEX ·   199

urban space, 124–126
and the mind, 135–146
  in biology, 138–139, 146
    biospace, 138–139, 146
  in chemistry, 136–137, 146
    chemospace, 137, 146
  in psychology, 141–147, 146
    psychospace, 141–142
and the post-human alteration
  of space-time, 165–171, 176
and the post-human challenge,
  164
and the post-human future,
  171–172
and the technological conquest
  of the micro-world, 175
definitions of, 3–4
in relation to engineering more
  dimensions of space-time,
  7–8, 19, 164
in relation to stretching/
  shrinking of space-time, 7–8,
  19, 164, 170
in relation to the foundation
  fallacy in the theoretical debate
  on space-time, 8, 19
in relation to the physical
  challenges to hyper-spatial
  consciousness, 177
theoretical debate on, 3–8, 18–19
see also Space, space-time, time
space colonization, 163–178
see also Space, space-time, time
space-time, 3–7
  and stretching/shrinking of
    space-time, 7–8, 19, 164, 170
  and engineering more
    dimensions of space-time,
    7–8, 19, 164
  and the foundation fallacy in
    the theoretical debate on
    space-time, 8, 19
  and the physical challenges to
    hyper-spatial
    consciousness, 177
  and the post-human alteration,
    165–171, 176
  and the post-human challenge,
    164
  and the post-human future,
    171–172
  and the technological conquest
    of the micro-world, 175
  definitions of, 3–4
  in relation to culture, 95–108
  in relation to nature, 149–160
  in relation to society, 111–132
  in relation to the mind, 135–146
  qualification of "better" in post-
    human forms of, 16–7
  theoretical debate on, 3–8, 18–19
  see also Space, space-time, time
spherical topology, 127–128, 137
  see also Space
spiritual
  and capitalist economics, 29–33
  and eco-feminist economics, 34
  and feudalist economics, 27
  and hunting/gathering
    economics, 27
  and Islamic economics, 35
  and Marxist economics, 34
  and mercantilist economics, 28
  and physiocratic economics, 28
  and post-capitalism, 36–43
  see also Post-human elitist
    calling, trans-feminine calling,
    trans-Islamic calling,
    trans-outerspace calling,
    trans-Sinitic calling
Steinhardt, Paul, 166
stewardizing process, the
  see Civilization
stretching
  and shrinking of space-time,
    7–8, 19, 164, 170
structural, the
  and civilizational holism, 86–91
  and floating consciousness, 25
  and the multiple causes of the
    emergence of post-capitalism,
    48–9

and the multiple causes of the
   emergence of post-democracy,
   58–59
in relation to space-time, 120–
   124, 132
see Methodological holism
sublime, the
   and existential dialectics, 9
   and the equality/inequality, 9,
      24
   and the freedom/unfreedom
      dialectics, 21
   in relation to
      moderntiy, 65–66
      postmodernity, 67–68
subliming process, the
   see Civilization
successive time, 106–106, 108
   see also Time
superfluid
   see Matter
systemic, the
   and civilizational holism, 86–91
   and floating consciousness, 26
   and the multiple causes of the
      emergence of post-capitalism,
      48–9
   and the multiple causes of the
      emergence of post-democracy,
      58–59
   in relation to space-time, 124–
      132
   see Methodological holism
symmetry
   and asymmetry, 12–3
   and the symmetry-asymmetry
      principle, 13, 19, 81–85
   see also Existential dialectics
symmetry-asymmetry principle, the,
   13, 19, 81–85
   see also Existential dialectics
system integration/fragmentation, 11

•T•

technological, the
   and existential dialectics, 9
   and the equality/inequality, 9,
      23
   and the freedom/unfreedom
      dialectics, 20
   and the micro-world, 175
   from pre-modernity to after-
      postmodernity, 70–71
   in relation to
      moderntiy, 65–66
      postmodernity, 67–68
technological time
   see Time
thinking-machines, 9–14
   see also Post-humans
thinking-robots, 9–14
   see also Post-humans
time, 3–8
   and culture, 95–96, 101–108
      linear and cyclic time,
         103–106, 108
      monochronic and polychronic
         time, 102–103, 108
      simultaneous and successive
         time, 106–106, 108
   and nature, 149–160
      in micro-physics, 152–153, 160
         quantum-mechanical time,
            153, 160
         shortest time, 152–153, 160
      in macro-physics, 156–158,
         160
         imaginary time, 156, 160
         reversible time, 156–158,
            160
   and society, 111–132
      in social organizations,
         114–116
            coordination time,
               115–116
            responsibility, 114–115
      in social institutions,
         118–120

· INDEX ·   201

monetary time, 119–120
real time, 118–119
working time, 120
in social structure,
    122–125
    cooperative/competitive
        time, 122–123
    discriminatory time,
        123–124
in social systems,
    128–131
    continuous time,
        128–129
    technological time,
        129–130
    ecosocial time, 130–131
and the mind, 135–146
in biology, 139–140, 146
    biotime, 139–140, 146
    ecobiotime, 140, 146
in chemistry, 137–138, 146
    chemotime, 137–138, 146
in psychology, 142–144, 146
    psychotime, 141–144, 146
        conscious, 142–143, 146
        unconscious, 143–144,
            146
and the post-human alteration
    of space-time, 165–171, 176
and the post-human challenge,
    164
and the post-human future,
    171–172
and the technological conquest
    of the micro-world, 175
definitions of, 3–4
in relation to engineering more
    dimensions of space-time,
    7–8, 19, 164
in relation to stretching/
    shrinking of space-time, 7–8,
    19, 164, 170
in relation to the foundation
    fallacy in the theoretical debate
    on space-time, 8, 19
in relation to the physical

challenges to hyper-spatial
    consciousness, 177
theoretical debate on, 3–8, 18–19
    see also Space, space-time, time
time illusion, 156–158
    see also Time
topology, 127–128, 138
    see also Space
trans-feminist calling, 36, 46, 52, 56
trans-Islamic calling, 38–39, 46, 52,
    56
translocation
    in relation to space, 151
    see also Space
trans-Outerspace calling, 40, 46, 52,
    56
trans-Sinitic calling, 37, 46, 52, 56
true, the
    and existential dialectics, 9
    and the equality/inequality, 9,
        24
    and the freedom/unfreedom
        dialectics, 21
    in relation to
        moderntiy, 65–66
        postmodernity, 67–68
Turok, Neil, 166

· U ·

unconscious
    in relation to time, 143–144
    see also Time
unfreedom
    see Existential dialectics
United States, the
    and the understanding of space,
        113
universe, 152–158
    and pocket universe, 166
    and the post-human alteration
        of space-time, 165–171
    and the post-human challenge,
        164
    and the post-human future,

171–172
    and types of super-civilization,
        173–174
    see Multiverse
urban space
    see Space
Utzon, Jorn, 125

•V•

value ideals
    and alternatives to capitalist
        value ideals, 34–5
    capitalist, 29–30
    pre-capitalist, 27–28, 63–4
van Slyke, Craig, 115
Versailles, 126–127
    see also Architecture, space
vertical and horizontal space, 98–99, 108
    see also Space
Vogel, Douglas, 115
von Burg, xiii, xv

•W•

Walker, Ian, 104
warp drive, 176
wealth
    and existential dialectics, 74
Weill, P., 119
Wells, Herbert George, 163
Western Europe
    and linear vs. cyclic time,
        103–106
    and monochronic vs.
        polychronic time, 102–103
Wheeler, John, 154
width
    and dimensions of space-time, 4
    see also Space, space-Time
working time, 120, 132
    see also Time

Wright, Frank Lloyd, 126

•X•

X and neologisms, 17
X1835
    see Matter

•Y•

Yates, Joanne, 119

•Z•

Zen
    in relation to space, 155–156

Books also by Peter Baofu

• *Beyond Nature and Nurture* (2006) •

• *Beyond Civilization to Post-Civilization* (2006) •

• *Beyond Capitalism to Post-Capitalism* (2005) •

• Volume 1: *Beyond Democracy to Post-Democracy* (2004) •

• Volume 2: *Beyond Democracy to Post-Democracy* (2004) •

• *The Future of Post-Human Consciousness* (2004) •

• *The Future of Capitalism and Democracy* (2002) •

• Volume 1: *The Future of Human Civilization* (2000) •

• Volume 2: *The Future of Human Civilization* (2000) •